3rd International Conference on Reliability Engineering (ICRE 2018)

IOP Conference Series: Materials Science and Engineering
Volume 575

Barcelona, Spain
24 – 26 November 2018

ISBN: 978-1-5108-9452-5
ISSN: 1757-8981

Printed from e-media with permission by:

Curran Associates, Inc.
57 Morehouse Lane
Red Hook, NY 12571

proceedings.com

Some format issues inherent in the e-media version may also appear in this print version.

Institute of Physics
Dirac House, Temple Back
Bristol BS1 6BE UK

Phone: 44 1 17 929 7481
Fax: 44 1 17 920 0979

techtracking@iop.org

Additional copies of this publication are available from:

Curran Associates, Inc.
57 Morehouse Lane
Red Hook, NY 12571 USA
Phone: 845-758-0400
Fax: 845-758-2633
Email: curran@proceedings.com
Web: www.proceedings.com

Table of contents

Volume 575

2018 3rd International Conference on Reliability Engineering

24–26 November 2018, Barcelona, Spain

Accepted papers received: 24 June 2019
Published online: 13 August 2019

Preface

2018 3rd International Conference on Reliability Engineering

2018 3rd International Conference on Reliability Engineering

Peer review statement

Papers

Reliability theory and application

Reliability study of U-type multistage interconnection networks

E Stergiou, D Liarokapis, E Glavas, C Angelis and S V Margariti.....1

Reliability of Loading Rate in Gait Analysis

V Abolins, K Nesenbergs and E Bernans.....10

Structural Reliability of Dapped End Beams with Different Reinforcement Layouts under Dynamic Loading

Z I Syed, M Kewalramani and E S Hejah.....15

Identification and correlation analysis for performance shaping factors in flight crew opcration

Lijing Wang, Yanlong Wang, Yingchun Chen and Dayong Dong.....21

Structural reliability and performance evaluation of symmetric and asymmetric multi-storey buildings subjected to lateral wind forces

M A Kewalramani, Z I Syed, M S Almustafa and M S Hassouna.....29

Quality engineering

Runtime Contracts Checker: Increasing Robustness of Component-Based Software Systems

M Illarramendi, L Etxeberria, X Elkorobarrutia and G Sagardui.....35

Medical Device Integrated System Accelerated Life Testing

Khaled Atua PhD and Saeid Baniasad PhD.....41

Evaluation of Energy Harvesting Technique for Security Module

Y Nozaki and M Yoshikawa.....46

Extended Order High Gain Observer Based Stabilization of 2 DOF Pan Tilt Platform for Aerial Imaging System

O Altaf, F M Malik, R A Azim, S Hussain and N Ahmad.....55

A Fault Inference Method under Uncertainty: Case Study on Crankshafts in Fracturing Pumps

X Zhang, L B Zhang and J Q Hu.....64

A Design Method of Two DOF Joint at the Base of Heavy-load Manipulator Based on Linear Driving

Zongmiao Dai, Peizhang Wu and Qingyuan Wang.....72

Vulnerability verification for illegal attack on secure CAN communication

K Nishida, Y Nozaki and M Yoshikawa.....81

Improving the Absolute Accuracy by Online Interpolation Technique of Industrial Robots

Darpan Patel, Lars Lienenlüke, Simon Storms, Christian Brecher and Johannes Wied.....88

Fault Detection and Diagnosis for a Nonlinear Hydraulic Actuator

Juan Felipe Ortiz, Carlos Borrás and Diego Sepulveda.....100

Research on Condition Monitoring of Key Components in wind Turbine based on Cointegration Analysis

Xide Sun, Dong Xue, Runfei Li, Xiaogang Li, Liye Cui, Xishan Zhang and Wei Wu.....112

An Intelligent Adaptive Robust Controller Based on Neural Network for a Pneumatic System

Ye Chen, Guoliang Tao and Fangan Meng.....119

Manufacturing Process Cost Reduction of Agricultural Machinery Part

Ramidayu Yousuk and Thidarat Rangsiearam.....126

Design engineering

Research on Antenna Cabin Structure Optimization Based on Response Surface and Multi-island Genetic Algorithm

Sun Weimin, Cao Weizhong and Gu Yeqing.....134

A Low Cost Robotic Medical Simulator for CPR Training

A A Freire Fiallos, F M Rosero Plúa and S N Rea Minango.....141

Study on a Real Time Scheduling Optimization Model for Multi-source Heterogeneous Distribution Network

Hongyang Huang, Xiaoming Huang, Boliang Lou, Junchao Ma, Haiqing Lu and Wentao Lv.....148

Finite Element Model Updating and Verification of Satellite-borne Slot Waveguide Based on Random Vibration Test

Sun Weimin, Cao Weizhong and Tong Mingbo.....157

PREFACE

On behalf of the organizing committee and technical program committee, we would like to extend to you a warm welcome to 2018 3rd International Conference on Reliability Engineering (ICRE 2018) in Barcelona, Spain.

The aim objective of ICRE 2018 is to provide a platform for researchers, engineers, academicians as well as industrial professionals from all over the world to present their research results and development activities in Reliability Engineering. This conference provides opportunities for the delegates to exchange new ideas and application experiences face to face, to establish business or research relations and to find global partners for future collaboration.

This year, we received 43 submissions and 21 papers are accepted for publication in conference proceedings. We continue to build on the theme of reliability engineering, with keynotes and papers on this important and growing topic. Every year at ICRE we strive to organize a conference that is technical strong while offering a great attendee experience. We look forward to your survey feedback in continuing to keep this conference a great experience in future years. Welcome to ICRE and I hope you will enjoy both the conference and the sunshine in Barcelona.

Prof. Michael Pecht
University of Maryland, USA

Conference Chair of ICRE2018

Conference Committee

International Advisory Committee

Prof. TANG Loon Ching, National University of Singapore, Singapore

Conference General Chair

Prof. Michael Pecht, University of Maryland, USA
Prof. Enrico Zio, Politecnico di Milano, Italy

Program Chair

Prof. Jin Wang, Liverpool John Moores University, UK
Prof. Francesco Di Maio, Politecnico di Milano, Italy
Prof. Yi Ding, Zhejiang University, China

Technical Committee

Prof. Antoine Rauzy, NTNU, Norway
Prof. Benoit IUNG, Lorraine University, France
Prof. Coen van Guijik, University of Huddersfield, UK
Prof. Daqing Li, Beihang University, China
Prof. David Coit, Rutgers University, USA
Prof. Detlef Schulz, Helmut-Schmidt University, Germany
Prof. Dr. Marko Cepin, University of Ljubljana, Slovenia
Prof. Emmanuel Fragniere, University of Applied Sciences Western Switzerland, Switzerland
Prof. Jiejuan Tong, Tsinghua Unversity, China
Prof. K. S. Trivedi, Duke University, Durham, North Carolina, USA
Prof. Khalifa Mansouri, University Hassan II Casablanca, Morocco
Prof. Liudong Xing, University of Massachusetts, USA
Prof. Mehmet Savsar, Kuwait University, Kuwait
Prof. Mitsuhiro Kimura, Hosei University, Japan
Prof. Moosung Jae, Hanynag University, Korea
Prof. Mustapha Nourelfath, Laval University, Canada
Prof. Pieter van Gelder, Technische Universiteit Delft, Netherlands
Prof. Radim Bris, Technical University of Ostrava, Czech Republic
Prof. Tse Peter, City University of Hong Kong, Hong Kong
Prof. Wei-Chang Yeh, National Tsing Hua University, Taiwan
Prof. Yanfu LI, Tsinghua University, China

IOP Conf. Series: Materials Science and Engineering **575** (2019) 011003 doi:10.1088/1757-899X/575/1/011003

Peer review statement

All papers published in this volume of *IOP Conference Series: Materials Science and Engineering* have been peer reviewed through processes administered by the proceedings Editors. Reviews were conducted by expert referees to the professional and scientific standards expected of a proceedings journal published by IOP Publishing.

IOP Conf. Series: Materials Science and Engineering **575** (2019) 012001 doi:10.1088/1757-899X/575/1/012001

Reliability study of U-type multistage interconnection networks

E Stergiou[1], D Liarokapis[1], E Glavas[1], C Angelis[1], S V Margariti[1]

[1]Department of Computer Engineering, TEI of Epirus, Arta, Greece

Corresponding E-mail: dili@teiep.gr

Abstract. Multistage Interconnection Networks (MINs) are used in wide applications of multiprocessor systems in order to set up connections between various nodes such as processors and memory modules. Reliability can be considered as one of the most challenging issues in these systems because this can degrade or cancel the network performance and consequently the effort of the whole system. Here, a type of multistage fabric called U-type is selected for study, particularly in terms of its reliability. This fabric turns out to be the most appropriate type for constructing parallel systems due to the particular features that it possesses. This study reveals that U-type fabrics are ideal devices as they present mostly high values of reliability and low values of latency compared to unidirectional fabrics. This behaviour becomes more intensive when it is used in a cluster that demands service that includes a high percentage of "local type traffic." Thus, U-type multistage devices can considerably enhance the connection points in constructing parallel systems, improving internal data flows.

1. Introduction

Multistage interconnection networks (MINs) are usually used as interconnection networks in cases of gigabit ethernet or Asynchronous transfer mode (ATM) switches [1-3]. But beyond this, MINs are proposed for connecting a large number of processors to establish a multiprocessor system [4-6]. Interconnection networks are critical in linking a large number of processors in the effort of building an integrated multiprocessor system. This is because such fabrics in general provide linear *bandwidth* and logarithmic packet latency in a number of nodes. Because of this behaviour, the interconnection networks are inevitable components in the construction of modern networks on chips (NoC) [5].

In constructing parallel systems, the most suitable type of MINs are the devices that here are named U-type MINs. U-type MINs can be considered as a special subclass of MINs. The basic and obvious feature that these devices have is that the inlets and outlets are located on the same side. This feature is extremely convenient for purely technical and constructional reasons when a large multiprocessor system is being manufactured. Due to this construction, the packets travel via a U-type device following a "U" shaped path. A second feature which these devices have is that in an arbitrary inlet–outlet pair, the packets do not follow the same route in terms of the number of hops. Ascertaining on the one hand the continuing need for such network interconnection systems and on the other hand the study gap that exists in the current research, we are attempting to study this subject in accordance with the latest constructional requirements. Similar approaches to clos class of interconnected networks have already taken place [7]. In brief, the novel contribution of this work can be summarized as:

- A special type of MIN called a U-type MIN was selected for study because this device is unavoidable when trying to build systems with multiple parallel processors.

- A study of the proposed fabric is carried out, showing that this fabric has many benefits; the results from the proposed fabric in terms of metrics, the *reliability* factor and packet *latency*, are extremely encouraging.
- The results yielded indicate that the U-type MINs outperform the corresponding (in terms of network size) *unidirectional* MINs in terms of the *reliability*, and *latency*.

The remainder of this paper is organized as follows: in the second section, the proposed U-type MIN is introduced, and details of its operation, internal paths and some key definitions are given. In the third section, a reliability analysis and relevant plots are shown and discussed. In the next section, numerical results relating to latency are given. Finally, in the Conclusion section, our conclusions and anticipated future work are presented.

2. Description of the U-type MIN

Figure 1 shows a typical view of a sub-system in a multiprocessor system. As can be seen in Figure 1, before the U-type MIN component a special case unit is used called an Adaptation Block (AB).

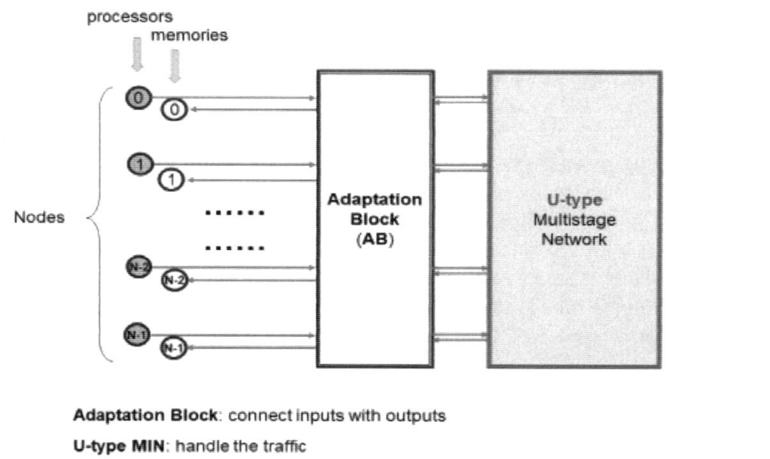

Figure 1. Typical view of a multiprocessor sub-system.

The NxNclassic MIN, where N is the number of inlets/outlets, consisting of $k \times k$ unidirectional switch elements (SEs) has a maximum number of stages $L = log_k N$, where k depicts the size of the SEs that are used to construct the fabric.

In U-type MINs, instead of the traditional *unidirectional* SEs with k input and k output ports, *bidirectional* $(2k \times 2k)$ SEs are used. Figure 2 depicts the typical $2k \times 2k$ SEs $(where\ k = 2, 4, 8, ...)$ that are used in the construction of U-type MINs. Depending on the set of *bidirectional* SEs that are selected, a corresponding U-type MIN is constructed. In general, *bidirectional* SEs have the ability to implement three ways of connecting, so they can operate in three modes. The first way of operation is the classic *forward* direction, the second is the *backward* direction and the third mode of operation is the return direction. (See Figure 2).

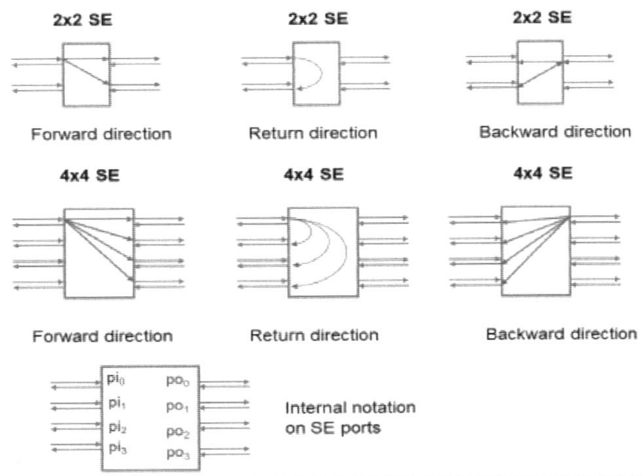

Figure 2. A typical operation (*forward*, *backward* and *return* operation) of *bidirectional* SEs that are used in U-type MINs.

In order to denote the working connection in a bidirectional SE, special notation is introduced, particularly:

- In the case of *forward* direction operation, the notation $pi_i(po_j)$ means connection between *input port* pi_i and *output port* po_j, where $i, j \in [0, k-1]$.
- In the case of *backward* direction operation, the notation $po_i(pi_j)$ is used, meaning connection between *output port* po_j and *input port* pi_i where $i, j \in [0, k-1]$.
- In *turnaround* direction operation the pair $[pi_i, pi_j]$ denotes the connection link between an input pi_i and output pi_j, where $i, j \in [0, k-1]$ and $i \neq j$.

In all cases the U-type MINs use the store and forward mechanism as a packet management technique.

2.1. Job cluster example

In Figure 3 an example of an arbitrary distribution of traffic load is shown. In this example, the microprocessors **0**, **1** are supposed to be running a job. This means the microprocessors feed the inputs **0** and **1** and the packets are driven to corresponding outputs **6** and **7**. This type of traffic is characterized as "*non-local traffic*," because the packet load is going through a long path (turn level 2 in MIN), and this has the effect of resulting in high values of packet delay.

On the other hand, the microprocessor **2** is assigned to execute another job and the packets are driven to output **3**. Because the distance between the input and output is the minimum (they are neighbours) the corresponding U-type path that the packets are using is the shortest. The same happens for the input/output pair **4**, **5**. This type of traffic is characterized as "*local traffic*." The packets which use the shortest path have the minimum corresponding packet *delay*.

The challenge here is how to minimize the packet path distances using the shortest U-type paths in order to improve the behaviour of the system by decreasing the time for job completion.

The main idea that is behind the usage of U-type MINs is to exploit the neighbouring inlets/outlets – exploit the shortest paths – in the interconnection network. Exploiting the shortest paths in the network reduces the packet traffic population in the U-type MIN and reduces traffic congestion. This characteristic basically decreases the *packet latency*, which is extremely useful in the case of Network on Chip (NoC) and generally in all attempts to redistribute traffic load and increase the performance of parallel computers.

2.2 Adaptation Block (AB)

The Adaptation Block (AB) accepts jobs by processors and has the ability to rearrange those jobs, changing the feed sequence at the inputs of a U-type MIN. The main goal of the AB is to increase the ratio of 'local type traffic' in the multistage network by changing the topology. The AB unit has the ability to rewire the connection between inputs/outputs. Thus, the AB has the ability to rearrange the input traffic in the U-type MIN. Each application allocates an exclusive subset of processors [8]. This bounded set of processors can be seen as a processor cluster.

Communication in a parallel multiprocessor system takes place predominantly inside these clusters. Therefore, it is desirable that processors within a cluster are placed as close to each other as possible. Ordinarily, allocating applications in adjacent processors is not feasible at runtime. Moreover, this situation gets even worse when the applications have different starting times relative to each other and when a favourable allocation of clusters is mostly not possible. The moving of an application in context from one processor to the other to localize the clusters is a very costly process [9]. The AB unit reconfigures the network topology to localize application clusters and optimize the communication requirements.

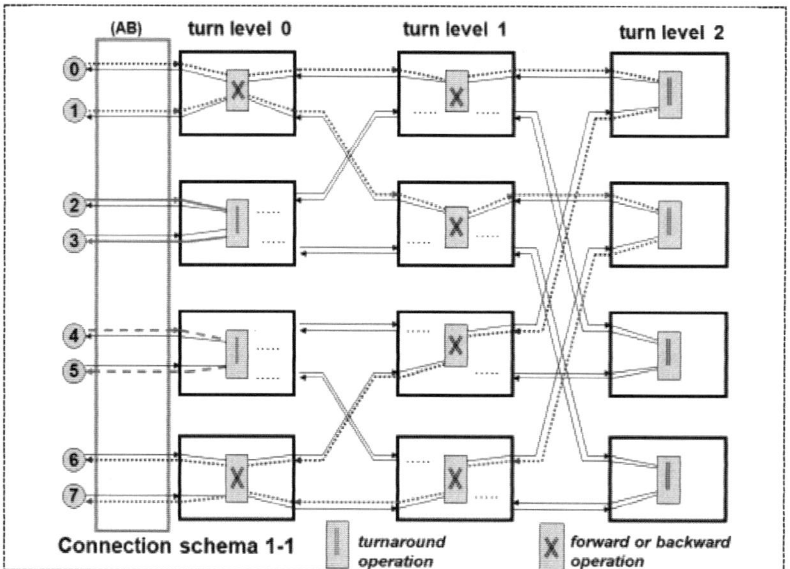

Figure 3. U-type MIN with arbitrary distribution traffic and with initial ingress connection schema 1-1 in the AB unit.

Supposing we have an aforementioned sub-system of a parallel system, a significant parameter for the proper operation of a multiprocessor system is the reliability of the system and especially the reliability of the U-type multistage fabric that is used in the system. Next, some basic definitions are presented followed by a description of the *reliability* analysis of the U-type MIN.

2.3 Basic Definitions

Buffer size (b): the maximum number of packets that an input buffer of an SE can hold.

Probability of packet arrivals (p) at an input (offered load): in our experiment the probability of packet arrivals is ranked from 0.1 to 1.0, with steps equal to 0.1.

Arrival process at the input of the U-type MIN: the generation of succeeding packets is assumed to be a mutually independent process. The arrival process for packets at the output queues of the first stage of the network $x_{k,n}^{(0)}$ is given by a binomial distribution $bin\left(k, \frac{p}{k}\right)$, where p is the fixed probability of a packet generated by a processor at each cycle. Therefore:

$$x_{k,n}^{(0)} = \begin{cases} \binom{k}{n} \cdot \left(\frac{p}{k}\right)^{n} \cdot \left(1-\frac{p}{k}\right)^{k-n} & , for\ 0 \le n \le k \\ 0, & , otherwise \end{cases} \quad (1)$$

3. Essential reliability analysis of U-type MINs

The *reliability* of a MIN is defined as its ability to overcome all unexpected circumstances. In this study, the point-to-point reliability (*p-t-p reliability*) is considered. This type of reliability is also known as the *terminal reliability*. With *p-t-p reliability*, the probability of at least one fault-free path existing between an inlet–outlet pair is calculated. The reliability between an arbitrary given pair (inlet–outlet) is dependent on all of the switch elements involved, insofar as the failure of each element implies the failure of the current routing action [10]-[13].

3.1 Reliability of the unidirectional NxN MINs

In the case of *unidirectional* NxN MINs, $L = log_k N$ isthe number of stages, where k depicts the size of the *unidirectional* SEs that are used to construct the fabric. In such a case, it is supposed that all of the packets traverse a series of L SEs components, so all the packets travel equal paths in terms of the node (or stage) hops. Assuming that each *unidirectional* SE has a reliability of r_i, then the reliability (R) of this subsystem can be calculated as

$$R_{UnidirectionaMIN} = \prod_{i=1}^{L=\log_k N} r_i \quad (2)$$

Supposing the *reliability* of all SEs are the same due to the homogeneity of the MINs, the total terminal reliability is $r^{log_k N}$.

3.2 Reliability of the U-type MIN

In the case of NxN *U-type* MINs, with a total depth $L = log_k N$ number of stages, the length of the paths that the packets use is not fixed in general. The path lengths (in number of node hops) of various packets can vary from 2 to $n = (2 \times \log_k N) - 1$. Two hops is the shortest U type path in a U-type MIN. This route is taken when the input and the output are adjacent and there is no other node involved between this pair. On the other hand, $n = (2 \times \log_k N) - 1$ is the longest path. The longest path happens when the distance between the input and the output is the maximum. So, the corresponding maximum value of the *terminal reliability* can be calculated as

$$R_{U-typeMIN} = \prod_{i=1}^{(2\times\log_k N)-1} r_i \ , \quad (3)$$

where r_i is the reliability of an arbitrary *bidirectional* SE acting in one of the three aforementioned operations. This reliability factor (r_i) is a manufacturer characteristic of the construction.

Consequently, the total *terminal reliability* value is in the range

$$R_{U-typeMIN} \in \left[\prod_{i=1}^{2} r_i \, , \prod_{i=1}^{(2\times log_k N)-1} r_i \right] \quad (4)$$

so, supposing the fabric has homogeneity, without harming generality the total terminal reliability value varies as

$$R_{U-typeMIN} \in [r^2 \text{-} r^{[(2\times\log_k N)-1]}], \text{ where } r = r_i \ \forall i \in [1 - ((2 \times log_k N) - 1)] \quad (5)$$

In Figure 4, the minimum values of the *normalized terminal reliability* for U-type MINs ($8x8$, $64x64$, $256x256$ and $1024x1024$ *network size*) are portrayed for varying values of the *bidirectional* SE's *reliability* (r_i), acting in one way. The minimum values of the total *normalized terminal reliability* occur when the packets travel the maximum route (in hops) inside the fabric. Thus, for U-type MINs with a network size of $1024x1024$ (MIN depth equal to 10) and SEs with *bidirectional* SE reliability of 90%, the total terminal reliability drops below 20%. Strictly speaking, that means this network size is prohibitive for such uses. According to this scenario, in all of the U-type MINs under study, all packets of traffic follow the longest "U" type route, in other words, the maximum route.

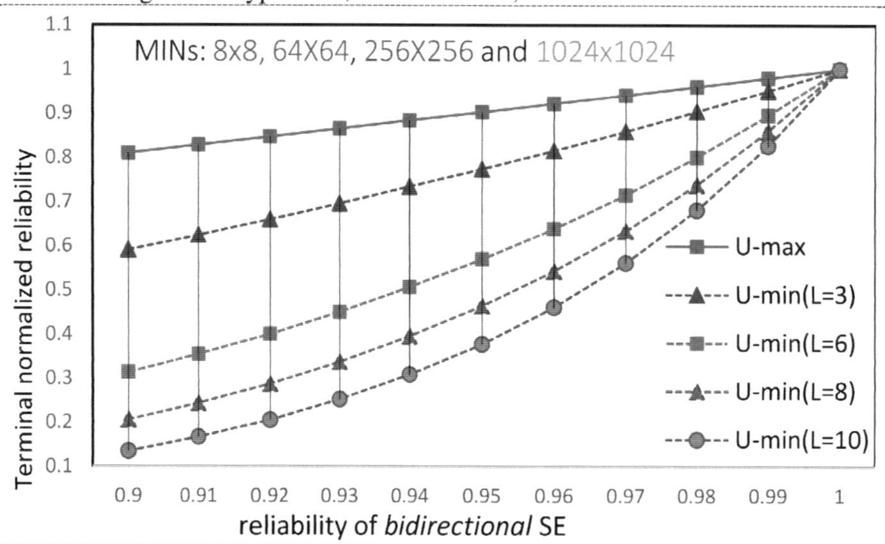

Figure 4. Terminal reliability fluctuation of U-type MINs with various network sizes (L=3, 6, 8 and 10) versus *reliability* (r_i) of the *bidirectional* SEs.

Also, the upper solid line in Figure 4 depicts the maximum values of the *normalized terminal reliability* of all the corresponding U-type MINs. The maximum values of the *terminal reliability* occur when the packets use the shortest path, equal to 2 hops, use only one *bidirectional* SE and operate exclusively in the *return* operation. The maximum set of *terminal reliability* values is common for all of the U-type MINs regardless of their size.

3.3 Traffic patterns of the cluster

Nonetheless, when we have a given cluster that includes various jobs, the packages of those jobs follow different paths in general according to the needs of each job (see Figure 3). This in turn changes the reliability factor of a U-type MIN. So, the results that are illustrated in Figure 4 show the boundaries of the *reliability* factor. However, for a given cluster with a specific traffic distribution, the *reliability* of its fabric needs calculation. In order to monitor the changes that the reliability factor undergoes, three types of typical traffic patterns are defined for study reasons.

Traffic pattern A: 90% non-local traffic and 10% uniform type traffic. Thus, according to this scenario 90% of the packet population turns in the last stage of the U-type MINs.

Traffic pattern B: Uniform distribution of "U" type paths. This means that we have a gradual return of packages at each stage.

Traffic pattern C: 90% local traffic and 10% uniform type traffic. Thus, according to this scenario 90% of the packet population turns in the first stage of the U-type MINs.

Figure 5 depicts the *probability of turn* for the three types of traffic patterns (A, B and C type) in a 5-stage depth U-type MIN.

IOP Conf. Series: Materials Science and Engineering **575** (2019) 012001 doi:10.1088/1757-899X/575/1/012001

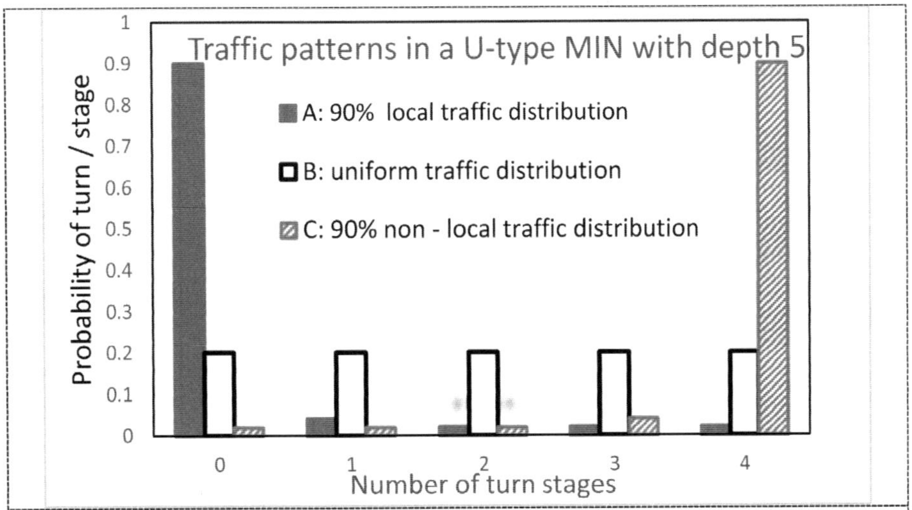

Figure 5. Probability of turn per stage for three types of traffic patterns (A, B and C) versus turn stage number in a U-type MIN.

3.4 Total terminal reliability for various traffic patterns

Figure 6 depicts the *total normalized reliability* for U-type MINs with depths varying from 3 to 10 stages and for the aforementioned three types of traffic patterns (A, B and C) in a quantitative manner. The upper solid line in Figure 6 shows the *total normalized reliability* when the traffic pattern is 90% *local traffic* for network depths varying from 3 to 10 stages. It is clear that the *total reliability* for this case remains at high values compared to the lower solid red line depicting the corresponding reliability of the opposite case (90% non-local traffic). In the case of *non-local traffic*, the reliability values drop dramatically.

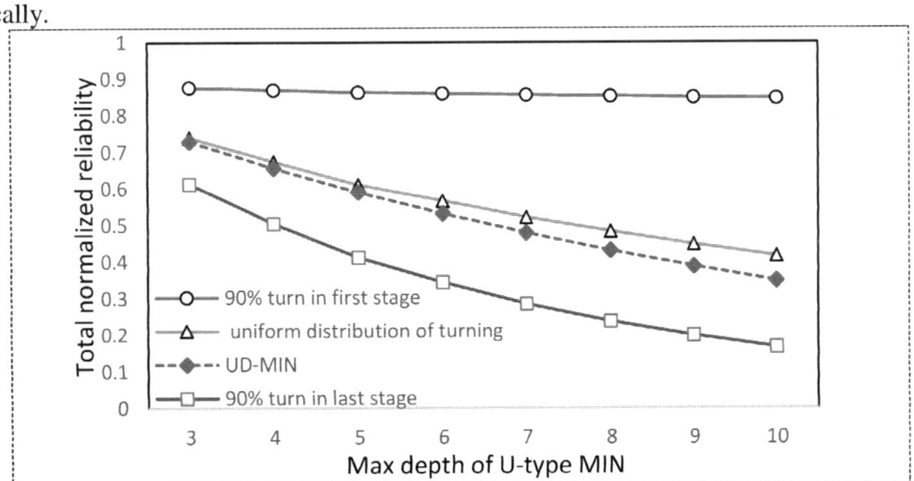

Figure 6. Total reliability for different turn patterns in various sizes of U-type MINs.

It's clear that the network size affects the values of the *total normalized reliability* of a U-type MIN. The prevalence of non-local traffic of *non-local traffic* in large-scale networks is the worst case, which should be avoided. Moreover, in Figure 6, the solid line with triangular triangular points illustrating the uniform distribution of turn traffic reveals that the values range over intermediate levels. Finally, the dashed line portrays the *normalized reliability* of a corresponding *unidirectional* MIN with an equivalent network size (5 stages, 32 × 32 MIN).

4. Latency metric of U-type MINs

Most of the studies of MIN performance for simplicity use simulation as an evaluation tool [14]. So, here for the same reasons, a special-purpose simulator was developed to evaluate the overall network performance of the U-type MINs. Also, another equivalent simulator was created for the *unidirectional* MINs. These tools were developed in C++ and are capable of operating under various configuration scenarios and handling various traffic patterns.

Figure 7 depicts the normalized packet *latency* of U-type and *unidirectional* (UD) MINs with the number of total stages equal to 6, 8 and 10 plotted versus the total offered load (varying from 10-100%). The solid lines depict the latency of *unidirectional* MINs while the dashed lines show the *latency* of U-type MINs. All of the MINs under study have a buffer size equal to 1, and the pattern applied to the U-type MIN inputs was 80% *non-local* traffic (Long Distance paths – LD) while the rest (20%) was uniform traffic.

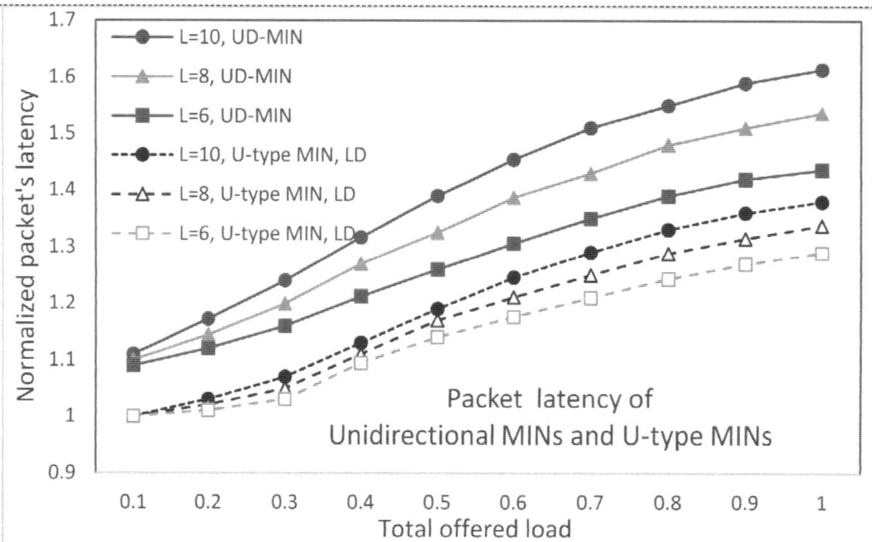

Figure 7. Normalized packet latency versus the total offered load for U-type and unidirectional MINs with maximum network depths of L=6, 8 and 10.

The plot reveals that the U-type MIN provides significantly less delay in the flow of packets in comparison to the corresponding (in terms of network size) *unidirectional* (UD) MIN. For example, an 8-stage (256x256) U-type MIN with full traffic (100% offered load of which 80% is *non-local traffic* (LD paths) and 20% is *uniform* traffic) presents a normalized packet latency equal to 1.33, whereas it is 1.53 for the corresponding *unidirectional* MIN (reduction of ~13%). This behaviour (*latency* reduction) is particularly useful in constructing parallel systems.

5. Conclusion

U-type MINs consisting of *bidirectional* SEs have been selected for study in terms of their *reliability* and packet *latency*. This class of MINs is particularly useful in constructing parallel systems. The current study shows that the *reliability* metric of U-type MINs is in general higher compared to *unidirectional* MINs. This metric is extremely good when a U-type MIN is fed by a large percentage of *local traffic*. This means that the jobs submitted in the system have a small distance between the input/output pair (e.g., they are neighbours). This preliminary study reveals that increased values of *reliability* combined with low packet *latency* values make U-type MINs an absolutely essential component in the construction of parallel systems.

U-type MINs are adroit multi-connection systems that have the ability to efficiently handle different types of traffic (e.g., hot spot, multicast, etc.) thus improving network performance. Therefore, this construction seems to have significant merit for use in the information industry. This work could be extended in many ways. For example, the system could be studied under hot spot traffic or multicast service and also in terms of performance metrics, priorities or cost metrics when operating with a backpressure mechanism.

6. References

[1] Soumiya T Nakamichi K Kakuma S Hatano T & Hakata A (1999). The large capacity ATM backbone switch "FETEX-150 ESP". *Computer Networks*, *31*(6), 603-615.

[2] Garofalakis J & Stergiou E (2011). Mechanisms and analysis for supporting multicast traffic by using multilayer multistage interconnection networks. *International Journal of Network Management*, *21*(2), 130-146.

[3] Ra'ed Y A & Mouftah H T (1995). Survey of ATM switch architectures. *Computer networks and ISDN systems*, *27*(12), 1567-1613.

[4] Abandah G A & Davidson E S (1996, April). Modeling the communication performance of the IBM SP2. In *Parallel Processing Symposium, 1996., Proceedings of IPPS'96, The 10th International* (pp. 249-257). IEEE.

[5] Dabiri B Rezaei S H S & Modarressi M (2017, August). Low-power Parallel Data Processing Using Computation Reuse. In *Proceedings of the 7th International Conference on Information Communication and Management* (pp. 1-4). ACM.

[6] Sadawarti H Bansal S & Kaur N (2015). A Survey on Multi-Stage Interconnection Networks. *Int. J. Comput. Sci. Trends Technol*, *3*(1), 143-151.

[7] Zhao Z Li B Hu Z Guan Y Wu B Dong X & Zhu G (2017, November). Minimizing delay by multi-hop routing and scheduling in clos networks for DCNs. In *Systems and Informatics (ICSAI), 2017 4th International Conference on* (pp. 1018-1023). IEEE.

[8] Liu J C Shin K G & Chang C C (1995). Prevention of congestion in packet-switched multistage interconnection networks. *IEEE Transactions on Parallel and Distributed Systems*, *6*(5), 535-541.

[9] Lu C & Lau S M (1994, August). A performance study on load balancing algorithms with task migration. In *TENCON'94. IEEE Region 10's Ninth Annual International Conference. Theme: Frontiers of Computer Technology. Proceedings of 1994*(pp. 357-364). IEEE.

[10] Jahanshahi M & Bistouni F (2015). Improving the reliability of the Benes network for use in large-scale systems. *Microelectronics Reliability*, *55*(3-4), 679-695.

[11] Rajkumar S & Goyal N K (2016). Review of multistage interconnection networks reliability and fault-tolerance. *IETE Technical Review*, *33*(3), 223-230.

[12] Md Yunus N A Othman M Mohd Hanapi Z & Lun K Y (2016). Reliability review of interconnection networks. *IETE Technical Review*, *33*(6), 596-606.

[13] Arya S & Singh N (2017). Reliability Analysis of Modified Irregular Augmented Shuffle Exchange Network (MIASEN). *International Journal of Computer Science Issues (IJCSI)*, *14*(3), 59.

[14] Stergiou E & Garofalakis J D (2011). A Simulation Study for Optimizing the Performance of Semi-layer Delta Networks. In *SIMULTECH* (pp. 257-265).

Reliability of Loading Rate in Gait Analysis

V Abolins[1,2], K Nesenbergs[1] and E Bernans[2]

[1] Institute of Electronics and Computer Science, Riga, Latvia
[2] Latvian Academy of Sport Education, Riga, Latvia

valters.abolins@edi.lv

Abstract. Loading rate is an important parameter of running gait as it is commonly associated with increased risk of injury. In this paper we show, that most of the research related to calculating loading rate values from ground reaction force data cannot be directly compared, as the absolute loading rate values and the related first impact peak location and height are highly dependent on the filtering method used, cut-off frequency selected as well as loading rate calculation method selected. The results are based on running trial data from non-forefoot striking long distance runners gathered by the authors.

1. Introduction

Loading rate (LR) is an important parameter of running gait because it is commonly associated with increased risk of injury [1–3]. On the other hand, there is some contradicting research indicating no connection between LR and risk of injury [4,5]. In this paper, the authors show, that an important reason for these conflicting results could be the differing methodologies for calculating LR used in the research literature.

LR together with multiple other running gait parameters (e.g. stance time, first impact) is usually calculated from ground reaction force (GRF) data. GRF is a metric commonly used in running gait analysis, and can be used to acquire running gait parameters - either by itself or in conjunction with other data sources (e.g. torque can be calculated from GRF and video data). When talking about LR, usually only the vertical component of GRF (vGRF) is used, thus all further references to GRF in this article are assumed to be the vertical ground reaction force component only if not explicitly stated differently.

By definition the calculation of LR is simple - the maximum ground reaction force (F_{max}) reached by the foot during the first contact is divided by the time (T_{max}) to reach this force starting from the beginning of the impact. In practice, the calculation is more complex and has several sources of errors. The first source of errors is finding the point in the impact phase, where the F_{max} is reached (as well as the corresponding time T_{max}). This point is usually referred to as the point of interest (POI). The GRF data of runners who are not forefoot strikers contains a distinct first spike with a maximum point called the first impact, which is used as POI for LR calculations. This first impact point is not found in the GRF data of forefoot strikers leading to multiple approached for calculating POI each potentially leading to different results. One approach to selecting POI in forefoot striker running GRF data is to find the first point where the slope of GRF curve changes [6,7]. Another approach used is to calculate the average POI of all non-forefoot strikers and use this value as POI for forefoot striker LR calculations [8]. This calculation method leads to varying results, as it is dependent on the specific experiment data, its acquisition, filtering methods, and other variables. For example, Liebermann et al. [8] calculate the average POI for non-forefoot strikers at 6.2% of stance time which can be used as

POI for forefoot strikers. Willy et al. calculate the same value as 13% [9] and Boyer et al. as 14% [10] while Goss and Gross settle on a range between 3% and 12% [11].
The second source of potential errors while calculating LR is the specific calculation method selected based on the identified POI. Because of mechanical and measurement noise as well as the shape of the impact curve the very beginning of the impact phase as well as close surroundings of POI are not regarded as good values for calculating LR. Common methods usually determine a starting (F_{start}) and ending (F_{end}) force for calculating the slope of LR, such that $0 < F_{start} < F_{end} < POI$. For example, $LR_{200N\text{-}90\%}$ is calculated as the slope of GRF starting from the point where it reaches the value of F_{start} = 200N, and ending at F_{end} = 90% of POI [8], while $LR_{20\%\text{-}80\%}$ is calculated as the GRF slope from F_{start} = 20% of POI till F_{end} = 80% of POI [12] (See figure 1).

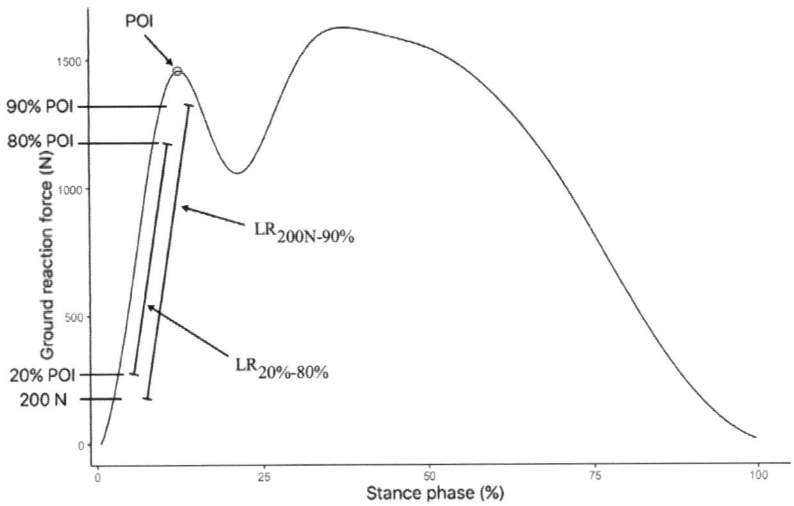

Figure 1. GRF curve example with POI and LR calculation ranges.

The third source of potential errors while calculating LR is the filtering method selected to process the GRF data before LR value is calculated. Even though there is no one filtering standard used for filtering running GRF data, a common choice is Butterworth low-pass filter, results of which depend on selected cut-off frequency. Researchers in their work use different cut-off frequencies, which can lead to different and incomparable results. Frequently used cut-off frequencies include 30 Hz [13–15], 50 Hz [16–18] and 100 Hz [19–21]. These frequencies are quite different and if used for pre-filtering the GRF can have a significant impact on the resulting LR. In this paper, we compare how calculated POI and resulting LR differ based on the choice of different filtering cut-off frequencies.

2. Methods

13 long distance non-forefoot strike runners (all male) participated in this study. The mean values ± SD of the key physical characteristics of the group are as follows: age = 26.85 ± 4.78 years, height = 180.54 ± 8.66 cm, weight = 69.92 ± 8.28 kg, time to run 21km = 72.5 ± 7.5 min. All participants were injury free 6 months prior to the experiment.
Running trials were conducted on a 30 m wooden track. GRF data was collected using two 600 mm x 400 mm force platforms (*BTS P-6000*, Italy), sampled at 1000 Hz. To control running speed, 3 photocells (*Microgate Polifemo Light Radio*, Italy) were used.
Running speed was set between 18 km/h and 19 km/h. Participants had time to practice and get used to this speed and running conditions. They could start the trials whenever they felt they are ready. They were asked to run in their preferred running shoes and not to try to hit the force platforms.
All participants completed 3 clean over ground running trials. During runs, GRF from force plates was measured. If there was no visual evidence of missing the force platforms and the speed was within 18-19 km/h range, the trial was accepted. Data of both feet stances was collected.

The trial for each participant which had the speed closest to 19 km/h was selected for further analysis. Both feet stances were analysed. GRF data was normalized in Body weight (BW) and was filtered using a low-pass fourth-order Butterworth filter with cut-off frequencies of 30 Hz, 50 Hz, and 100 Hz. POI, $LR_{200N\text{-}90\%}$ and $LR_{20\%\text{-}80\%}$ values were calculated from all filtered data. The stance phase was defined between ground contact (vGRF > 20 N) and take-off (vGRF < 20 N) [22].
Pearson's correlation coefficient was calculated for the POI, $LR_{200N\text{-}90\%}$ and $LR_{20\%\text{-}80\%}$ between each pair of cut-off frequencies. The significance of mean differences was calculated using Student's t-test with paired samples. All calculation was made in *R version 3.4.3.* [23].

3. Results

The mean POI was 16.87% (30 Hz), 13.39% (50Hz) and 11.54% (100Hz) for the respective cut-off frequencies. The POI position was decreasing when choosing higher cut-off frequency while resulting $LR_{200N\text{-}90\%}$ and $LR_{20\%\text{-}80\%}$ was increasing with higher cut-off frequencies (see table 1).

Table 1. LR and POI results with different cut-off frequencies.

	30 Hz *Mean (SD)*	50 Hz *Mean (SD)*	100 Hz *Mean (SD)*
POI (%)	16.87 (6.01)	13.39 (3.49)	11.54 (3.65)
$LR_{200N\text{-}90\%}$ (BW/s)	119.18 (21.76)	166.27 (50.52)	200.11 (63.73)
$LR_{20\%\text{-}80\%}$ (BW/s)	129.35 (22.45)	181.01 (56.64)	230.00 (77.74)

A significant difference in POI and both LR variation values between all three cut-off frequency combinations (see table 2) was observed. With higher cut-off frequency LR was higher.

Table 2. Differences of LR and POI results between pairs of cut-off frequencies.

		Mean	*95% CI*	*p-value*
30 Hz vs 50 Hz	**POI (%)**	3.48	(1.37, 5.58)	0.002
	$LR_{200N\text{-}90\%}$ (BW/s)	-47.09	(-59.62, 5.58)	< 0.001
	$LR_{20\%\text{-}80\%}$ (BW/s)	-51.66	(-66.20, -37.12)	< 0.001
30 Hz vs 100 Hz	**POI (%)**	5.33	(2.58, 8.08)	<0.001
	$LR_{200N\text{-}90\%}$ (BW/s)	-80.94	(-100.00, -61.87)	< 0.001
	$LR_{20\%\text{-}80\%}$ (BW/s)	-100.66	(-124.80, -76.51)	< 0.001
50 Hz vs 100 Hz	**POI (%)**	1.85	(0.05, 3.65)	0.045
	$LR_{200N\text{-}90\%}$ (BW/s)	-33.85	(-46.87, -20.82)	< 0.001
	$LR_{20\%\text{-}80\%}$ (BW/s)	-48.99	(-63.74, -34.25)	< 0.001

There was also a significant difference between $LR_{200N\text{-}90\%}$ and $LR_{20\%\text{-}80\%}$ when choosing different cut-off frequencies. When GRF was filtered with 30 Hz cut-off frequency mean difference between $LR_{200N\text{-}90\%}$ and $LR_{20\%\text{-}80\%}$ was -10.17 BW/s ($p < 0.001$), but when GRF was filtered with 50 and 100 Hz frequencies, the same difference was -14.72 BW/s ($p < 0.001$) and -29.89 BW/s ($p < 0.001$) respectively. $LR_{200N\text{-}90\%}$ was always lower than $LR_{20\%\text{-}80\%}$ (see table 3).

Table 3. Differences and correlation between two types of LR results at different cut-off frequencies.

	30 Hz	50 Hz	100 Hz
Mean difference (BW/s)	-10.17	-14.74	-29.89
95% CI	(-11.19, -9.15)	(-17.52, -11.97)	(-39.57, -20.20)
r (correlation coefficient)	0.994	0.998	0.962
95% CI	(0.986, 0.997)	(0.996, 0.999)	(0.915, 0.983)

There was a significant correlation in both loading rates between all 3 cut-off frequencies (Table 4). There was a significant correlation in POI between 30 Hz vs. 50 Hz cut-off frequencies, but there was no correlation between 30 Hz vs. 100 Hz and 50 Hz vs. 100 Hz cut-off frequencies (Table 4).

Table 4. Correlation of LR and POI results between pairs of cut-off frequencies.

		r (correlation c.)	*95% CI*	*p-value*
30 Hz vs 50 Hz	**POI**	0.505	(0.146, 0.746)	0.009
	$LR_{200N\text{-}90\%}$	0.941	(0.872, 0.974)	< 0.001
	$LR_{20\%\text{-}80\%}$	0.950	(0.891, 0.978)	< 0.001
30 Hz vs 100 Hz	**POI**	0.069	(-0.327, 0.444)	0.738
	$LR_{200N\text{-}90\%}$	0.832	(0.656, 0.922)	< 0.001
	$LR_{20\%\text{-}80\%}$	0.852	(0.693, 0.932)	< 0.001
50 Hz vs 100 Hz	**POI**	0.220	(-0.183, 0.560)	0.280
	$LR_{200N\text{-}90\%}$	0.865	(0.719, 0.938)	< 0.001
	$LR_{20\%\text{-}80\%}$	0.899	(0.786, 0.954)	< 0.001

4. Discussion

With a relative increase in the cut-off frequency used, the POI is reduced moving towards the beginning of the impact phase, while the calculated LR is increasing. In addition to the reduction of POI time, it has been noted by Kiernan et al., that force of the first impact (at POI) is increased if the cut-off frequency is increased [24]. Both of these factors contribute to the increase of LR, as it is a division of the force vs time at POI. Even though previous research states that average POI is located at fixed points our results show, that it is dependent on the cut-off frequency selected. For example, Lieberman et al. did not filter GRF data and arrived at 6.2% POI [8], while Willy et al. and Boyer et al. used Butterworth low-pass filter with 50Hz cut-off frequency arriving at respective POI values of 13% [9] and 14% [10]. The results acquired by the authors confirm this, showing that higher cut-off frequency leads to lower POI values and provides a similar 13.39% POI value at 50Hz cut-off frequency. The differences between both types of LR calculations are explained by the fact, that GRF data growth rate decreases as it approaches the POI, where growth rate is 0. Thus $LR_{200N\text{-}90\%}$ is lower than $LR_{20\%\text{-}80\%}$, because it covers 10% more of the GRF data, with a lower growth rate (see figure 2), leading to another reason why results of both of these methods cannot be directly compared in different research.

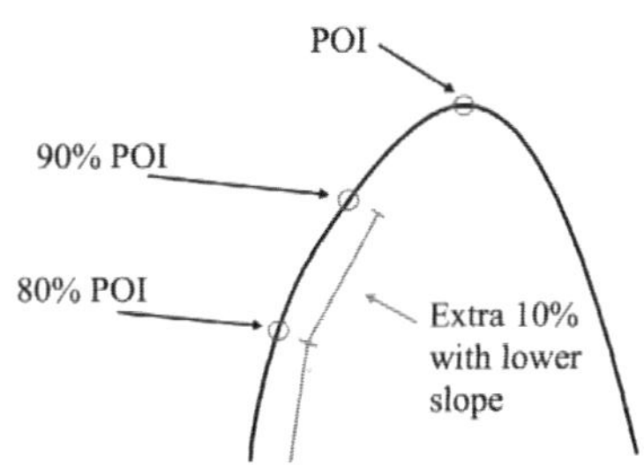

Figure 2. Locations of 80%, 90% and 100% POI on GRF curve.

For a specific case, even though absolute difference increases with the increased cut-off frequency it is not important which specific method of calculation is chosen, as they have high correlation and the ranging is preserved. The same relative ranging property is also preserved while one LR calculation method is used with different cut-off frequencies even though absolute values change.

This means, that for a specific case the choice of filtering cut-off frequency does not matter, as the relative comparison of LR values will hold, but POI and LR values cannot be compared with the work of other authors if their selected cut-off frequencies and trial running speeds are different, as the first impact peak dependence on the cut-off frequency also changes with running speed [24].

5. References

[1] Zadpoor A A and Nikooyan A A 2011 The relationship between lower-extremity stress fractures and the ground reaction force: A systematic review *Clin. Biomech.* **26** 23–8

[2] Grimston S K, Nigg B M, Fisher V and Ajemian S V. 1994 External loads throughout a 45 minute run in stress fracture and non-stress fracture runners *J. Biomech.* **27** 668

[3] Hreljac A, Marshall R N and Hume P A 2000 Evaluation of lower extremity overuse injury potential in runners. *Med. Sci. Sports Exerc.* **32** 1635–41

[4] Azevedo L B, Lambert M I, Vaughan C L, O'Connor C M and Schwellnus M P 2009 Biomechanical variables associated with Achilles tendinopathy in runners *Br. J. Sports Med.* **43** 288–92

[5] Messier S P, Edwards D G, Martin D F, Lowery R B, Cannon D W, James M K, Curl W W, Read H M and Hunter D M 1995 Etiology of iliotibial band friction syndrome in distance runners. *Med. Sci. Sports Exerc.* **27** 951–60

[6] Kluitenberg B, Bredeweg S W, Zijlstra S, Zijlstra W and Buist I 2012 Comparison of vertical ground reaction forces during overground and treadmill running. A validation study *BMC Musculoskelet. Disord.* **13** 235

[7] Laughton C A, Davis I M and Hamill J 2003 Effect of Strike Pattern and Orthotic Intervention on Tibial Shock during Running *J. Appl. Biomech.* **19** 153–68

[8] Lieberman D E, Venkadesan M, Werbel W A, Daoud A I, D'Andrea S, Davis I S, Mang'Eni R O and Pitsiladis Y 2010 Foot strike patterns and collision forces in habitually barefoot versus shod runners *Nature* **463** 531–5

[9] Willy R, Pohl M B and Davis I S 2008 Calculation of vertical load rates in the absence of vertical impact peaks *American Society of Biomechanics Meeting*

[10] Boyer E R, Rooney B D and Derrick T R 2014 Rearfoot and Midfoot or Forefoot Impacts in Habitually Shod Runners *Med. Sci. Sport. Exerc.* **46** 1384–91

[11] Goss D L and Gross M T 2013 A Comparison of Negative Joint Work and Vertical Ground Reaction Force Loading Rates in Chi Runners and Rearfoot-Striking Runners *J. Orthop. Sport. Phys. Ther.* **43** 685–92

[12] Williams D S, McClay I S and Manal K T 2000 Lower Extremity Mechanics in Runners with a Converted Forefoot Strike Pattern *J. Appl. Biomech.* **16** 210–8

[13] Jewell C, Boyer K A and Hamill J 2017 Do footfall patterns in forefoot runners change over an exhaustive run? *J. Sports Sci.* **35** 74–80

[14] Bullimore S R and Burn J F 2007 Ability of the planar spring–mass model to predict mechanical parameters in running humans *J. Theor. Biol.* **248** 686–95

[15] Riley P O, Dicharry J, Franz J, Croce U DELLA, Wilder R P and Kerrigan D C 2008 A Kinematics and Kinetic Comparison of Overground and Treadmill Running *Med. Sci. Sport. Exerc.* **40** 1093–100

[16] Bonacci J, Saunders P U, Hicks A, Rantalainen T, Vicenzino B (Guglielmo) T and Spratford W 2013 Running in a minimalist and lightweight shoe is not the same as running barefoot: a biomechanical study *Br. J. Sports Med.* **47** 387–92

[17] Bonacci J, Vicenzino B, Spratford W and Collins P 2014 Take your shoes off to reduce patellofemoral joint stress during running *Br. J. Sports Med.* **48** 425–8

[18] Graf E, Wannop J W, Schlarb H and Stefanyshyn D 2017 Effect of torsional stiffness on biomechanical variables of the lower extremity during running *Footwear Sci.* **9** 1–8

[19] Stefanyshyn D J and Nigg B M 1997 Mechanical energy contribution of the metatarsophalangeal joint to running and sprinting *J. Biomech.* **30** 1081–5

[20] Christina K A, White S C and Gilchrist L A 2001 Effect of localized muscle fatigue on vertical ground reaction forces and ankle joint motion during running *Hum. Mov. Sci.* **20** 257–76

[21] Santos-Concejero J, Tam N, Coetzee D R, Oliván J, Noakes T D and Tucker R 2016 Are gait characteristics and ground reaction forces related to energy cost of running in elite Kenyan runners? *J. Sports Sci.* **35** 1–8

[22] Nordin A D, Dufek J S and Mercer J A 2017 Three-dimensional impact kinetics with foot-strike manipulations during running *J. Sport Heal. Sci.* **6** 489–97

[23] R Core Team 2017 R: A Language and Environment for Statistical Computing

[24] Kiernan D, Miller R H, Baum B S, Kwon H J and Shim J K 2017 Amputee locomotion: Frequency content of prosthetic vs. intact limb vertical ground reaction forces during running and the effects of filter cut-off frequency *J. Biomech.* **60** 248–52

IOP Conf. Series: Materials Science and Engineering **575** (2019) 012003 doi:10.1088/1757-899X/575/1/012003

Structural Reliability of Dapped End Beams with Different Reinforcement Layouts under Dynamic Loading

Z I Syed*, M Kewalramani and **E S Hejah**

College of Engineering, Abu Dhabi University, UAE

*zubair.syed@adu.ac.ae

Abstract. Reinforced concrete Dapped End Beams (DEB), also known as half-joints, are used in bridges and many other pre-cast constructions to reduce end depth and increase lateral stability. Dapped end beams are expected to experience dynamic loads when used in bridges, the available past studies on the behavior and damage assessment of DEBs are mainly for static loading. The reinforcement layouts of DEBs can influence the behavior of these shear critical members under impact loading. Better understanding of the crack propagation and failure patter of DEBs under dynamics loads is required for safe and economic design of these structural elements. A non-linear numerical transient study was conducted to investigate the dynamic performance of DEBs with different reinforcement layouts. Advanced material models capable of including strain-rate effect and material non-linearity to capture realistic behavior of DEBs were used. The simulated models in finite element package LS-DYNA were verified and used to conduct detailed parametric study to investigate the impact behavior of DEBs with different reinforcement layouts. Sensitivity of concrete compressive strength, main dapped end reinforcement and special shear reinforcement detailing on the structural reliability of DEBs were studied.

1. Introduction

Dapped end beams are considered as shear critical structural elements due to the unusual shape at the beam ends where special design consideration is required due to sever stress flow disturbance [1]. Many studies have been completed on the behavior of DEBs when subjected to static loads [2-5]. In those studies, the main parameters observed were concrete strength, reinforcement layouts and cross-sectional dimensions of the beams. Under the applied static load, diagonal cracks will develop at the re-entrant corner, and as load increases, diagonal cracks will propagate towards the point of loading [4]. Cross-sectional dimensions played an important role in controlling the diagonal cracks [4]. Similarly, Aswin *et al.* [5] showed that higher dapped-end height provides higher failure load capacity. Moreover, concrete strength also has an effect on the shear strength of DEBs was shown in a study completed by Lu *et al.* [2], where the performance of dapped-end beams using high strength concrete showed that shear strength of DEBs increases with the increase of concrete compressive strength. The effect of reinforcement of dapped-end beam on its capacity and performance under static load has been also observed. Increasing the reinforcement ratio will enhance the strength capacity of these beams [5]. Minimizing the spacing between stirrups and increasing number of stirrups enhance the strength capacity of dapped-end beams under static load [3,4]. Similarly, inclined stirrups and bent reinforcement also were proven to increase the load capacity of dapped end beams when subjected to static loads [3, 4]. A recent study observed the effect of reinforcement layouts on the static behavior of DEBs [6]. Variation between diagonal, shear and horizontal reinforcement was done in the four beams in order to observe their effect on the load carrying capacity of dapped end beams under static loads. The study showed that under static load, the greatest impact on the failure load was seen in the beam without diagonal reinforcing bars. Beams with no diagonal or bent bars experienced failure in the nib area. Findings of this study [6] showed that reinforcement layouts have significant impact on the strength capacity of dapped-end beams.

Although, dapped-end beams are often exposed to dynamic loading, the dynamic behavior of dapped-end beams is not well established yet. In recent years, a growing interest has been observed in the field of structural behavior and response under dynamic loading which may involve large displacements, material non-linearity, elastic and plastic instability and material behavior under high strain rates. This research presents a study on dapped end beams for different layouts of main and hanger reinforcements for different concrete grades. Main parameters included in the study are concrete compressive strength, detailing of main dapped end reinforcement and special shear reinforcement (hanger reinforcement).

2. Finite Element Modeling

The finite element analysis software LS-DYNA was used in was used in performing the structural response predictions. Many studies have been completed on the behavior of rectangular beams under dynamic loading [7-10]. Verification of the modeling techniques and selected material models was done by modeling a previously tested rectangular beam by Zhan *et al.* [7]. All the modeling approaches and material models were verified by comparing simulation results with experimental results for rectangular beams. Once verified, same modeling techniques and approaches were followed in simulating DEBs. Due considerations were taken in element modeling, contact algorithms, material modeling, boundary conditions and load application. Detailed finite element modeling procedure is presented in a separate study [11]. All the properties related to the previously experimented beam were taken from Lu *et al.* [2]. The overall dimensions and reinforcement layout of the selected beam are shown in Figures 1 and 2.

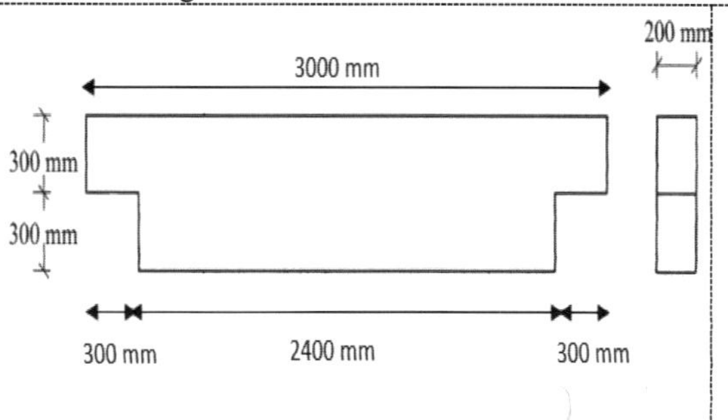

Figure 1. Dapped end beam configuration.

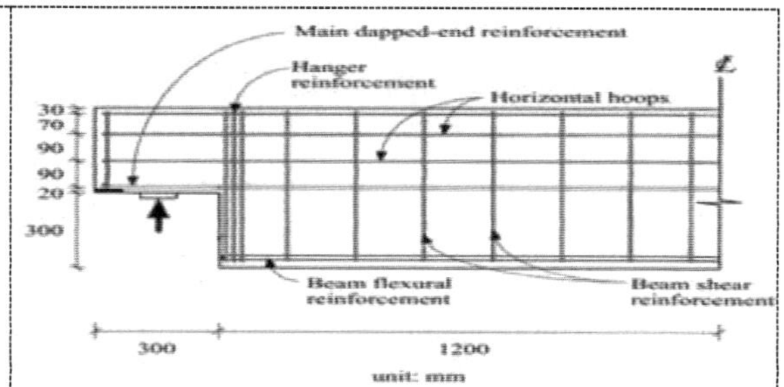

Figure 2. Reinforcement layout of modelled beam.

In case of modelling reinforced concrete beams, a perfect bond was assumed between steel and concrete elements and was achieved by assuring steel and concrete elements share common nodes in the discrete model. Mesh density is considered as one of the most important aspects of finite element modelling as due to strain gradient across an element, coarse meshing in complex areas of a structure may produce unreliable results. In modeling dapped-end beams, different meshing sizes were used in different locations, finer meshes were used in areas where higher stresses are expected to develop. In LS-DYNA, the most popular choice for modelling concrete in three dimensions is to use eight nodded hexahedral solid elements which has been found to be very efficient [12] are used in this study. Reinforcement steel and stirrups were simulated using beam and truss elements and all boundary condition parts were modeled using solid elements. To simulate behavior of reinforcement steel material in this study *MAT_PLASTIC_KINEMATIC (MAT_003) was used. For steel plates and cylinders used in modeling roller supports for the selected simply supported beam in addition to the steel impactor which represents the weight dropped as the external applied load was modelled as rigid

IOP Conf. Series: Materials Science and Engineering **575** (2019) 012003 doi:10.1088/1757-899X/575/1/012003

material. Rigid elements are bypassed in the element processing where no storage is allocated for storing history variables; thus, the rigid material type is very cost efficient [13].

In order to capture the complex nature of concrete, different material models were used for simulating its behavior and Concrete Damage Rel.3 (Material 72R3) was found to be able to capture the behavior of the beam closely to the experimental behavior. The Hexahedron solid elements used in concrete modeling are considered as under-integrated elements with only one integration point, therefore defining hourglass model is needed in order to resist any undesirable hourglass modes. To control hourglass, a stiffness hourglass formulation was used. Stiffness forms generate hourglass forces proportional to components of nodal displacement contributing to hourglass modes. Different hourglass types and coefficients were used through different stages of simulations in order to decide the most suitable ones for the simulated beam as using inappropriate hourglass type or coefficient can be noticed on the results. Previous analyses and studies suggest that hourglass energy should not exceed 10% of the internal energy [12] for both entire system and each part of the system. In case of the whole system, internal energy was 17,442 J and hourglass energy was 593 J, hence, the hourglass energy was well below 10% of the internal energy. While, in case of the energy of concrete material part only, the hourglass energy was 581 J while internal energy 16,190 J which also shows it was satisfying meeting the condition. On the other hand, deflection time histories obtained from the experimental tests on rectangular beam under impact load and from simulation of the same beam using the previously elaborated techniques are given in Figure 3. The figure demonstrates that the modelling techniques, selected material models and boundary conditions have produced results significantly close to the experimental values.

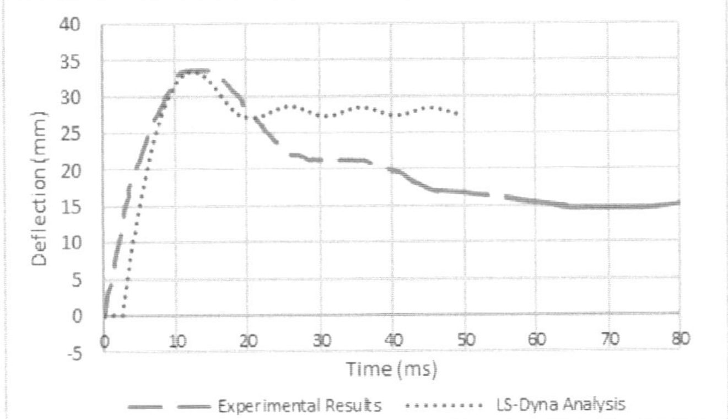

Figure 3. Deflection time-history comparison for model verification.

3. Finite Element Modeling

Main parameters tested in this study included main- and hanger-reinforcement layouts with different grades of concrete. This study aims to produce better understanding of the influence of various reinforcement layouts on the impact behavior of dapped-end beams. Reinforcement layouts varied in terms of the number of bars, diameter of bars and spacing between steel bars. Figure 4 shows different main dapped end reinforcement layouts, some reinforcement layouts had a very similar area of steel as the control beam and other had different ones. A total of 31 beam models were simulated in this research, all listed in Table 1.

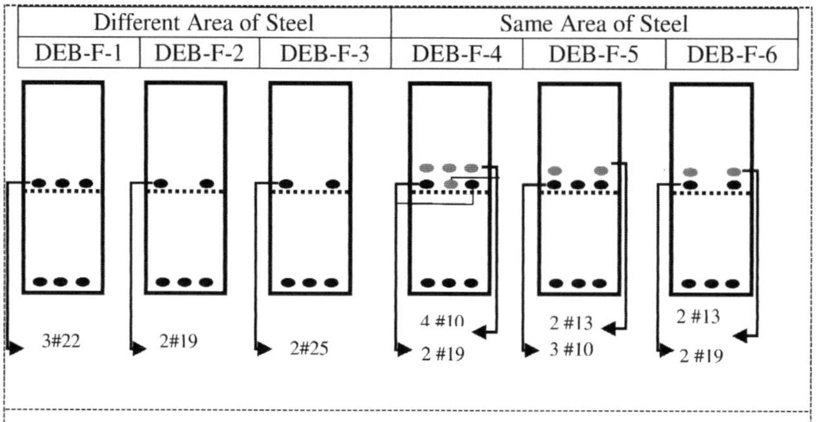

Figure 4. Simulated dapped-end main reinforcement layouts

Table 1. Symbols of dapped-end beam models along with all parameters variation

Main Parameter	Different Layout Notation		Model Symbol
Shear hanger reinforcement	3#10 @20 mm		DEB-S-1
	3#13 @20 mm		DEB-S-2
	5#10 @20 mm		DEB-S-3
	5#13 @20 mm		DEB-S-4
	2#13 @40 mm		DEB-S-5
	3#10 @40 mm		DEB-S-6
	3#16 @40 mm		DEB-S-7
Main dapped end flexural reinforcement	Different area of steel	3#22 (one layer)	DEB-F-1
		2#19 (one layer)	DEB-F-2
		2#25 (one layer)	DEB-F-3
	Same area of steel as control	2#19 and 4#10 (6 bars in two layers)	DEB-F-4
		3#16 and 2#13 (5 bars in two layers)	DEB-F-5
		2#19 and 2#13 (4 bars in two layers)	DEB-F-6
Concrete compressive strength and main dapped end reinforcement	3#22 (one layer)	f'c = 28 MPa, 40 MPa and 75 MPa	DEB-FcF-1a, 1b, 1c
	2#19 (one layer)		DEB-FcF-2a, 2b, 2c
	2#25 (one layer)		DEB-FcF-3a, 3b, 3c
	2#19 and 4#10 (6 bars in two layers)		DEB-FcF-4a, 4b, 4c
	3#16 and 2#13 (5 bars in two layers)		DEB-FcF-5a, 5b, 5c
	2#19 and 2#13 (4 bars in two layers)		DEB-FcF-6a, 6b, 6c

4. Results and Discussion

Concrete material model *MAT_CONCRETE_DAMAGE_REL3 doesn't have the property of showing the crack propagation in the concrete. Thus, an additional material can be defined in order to show the damaged elements due to the applied load. *MAT_ADD_EROSION was defined and linked to the concrete material. Accordingly, some damaged beam models were viewed. Viewing the plastic strain contours helped in showing the expected crack pattern in the concrete Figure 5 shows the plastic strain contour for compressive strength of 40 MPa. According to plastic strain contours shown in Figure 5, it is clear that initial cracks will start to develop in the reentrant area nearer to the application of load and more concentrated in the upper area on the right side of the applied load. The area of continuance plastic strain in the concrete is located in the critical location where stress concentration is developed. Contours of plastic strain clearly indicted that the dapped-end beam has a shear failure mode. The formation of cracks for the control beam is shown in Figure 6. The crack patters also confirm the plastic strain distribution and failure initiation in the beam.

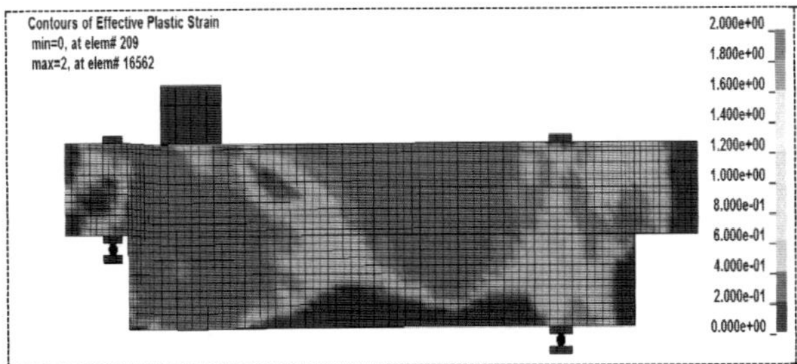

Figure 5. Contour of plastic strain under impact loading.

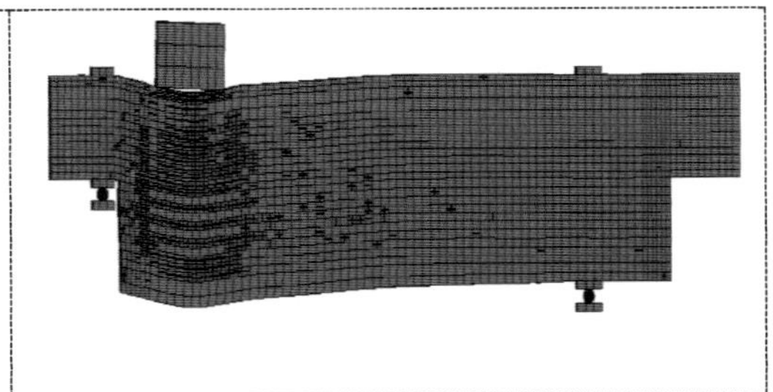

Figure 6. Crack pattern of the control beam obtained using erosion technique

The maximum shear stress was observed in the critical area of the dapped-end beam as expected. Generally, a very slight difference in shear stresses was noticed as detailing of both hanger and main dapped end reinforcements. Hanger reinforcement detailing didn't have much effect on the shear stress of high strength dapped end beams. The diameter of hanger reinforcement increased the shear strength of the high strength dapped end beam more than other factors of hanger reinforcement detailing did. On the other hand, concrete compressive strength had a greater effect on the shear strength of the dapped end beam under the applied impact load as shown in Figure 7. This agrees with the conclusion drawn by Lu *et al.* [2] that under static loads the shear strength of dapped end beams increases with the increase of concrete strength.

Stress-strain diagrams of main steel element for beam with different concrete compressive strengths are shown in Figure 8. When using a low strength concrete (28 and 40 MPa), steel showed a much rigid behavior than when compared to using a high strength concrete (70 and 75 MPa). This explains the fact that when lower strength concrete is used, steel is taking most of the load, while for higher grade of concrete, a large share of the load is handled by concrete.

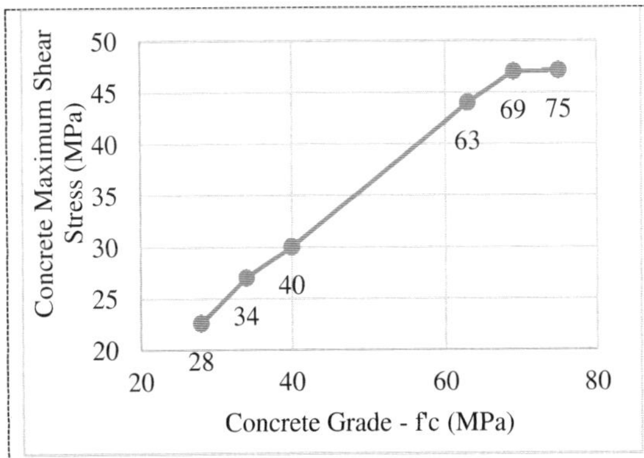

Figure 7. Effect of concrete compressive strength on the maximum shear stress.

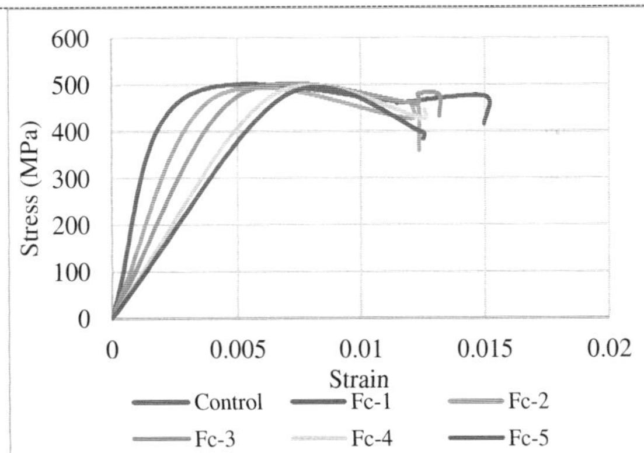

Figure 8. Stress-strain diagrams of main dapped end steel element for different compressive strength changes

The detailing of hanger reinforcement was found to affect the rigidity of the main dapped end steel insignificantly. In all models' steel reached the yielding point when almost all of them had the same maximum value of stress under the same applied load. Moreover, in all cases steel experienced almost the same rigidity. Under the same applied load, compressive strength of concrete and detailing of main dapped end reinforcement steel had the major effect on the rigidity of main dapped end steel elements. Lesser number of steel bars produced overstressed steel in case of DEB-F-2 and DEB-F-3 compared to other beams which indicates a higher chance of failure occurrence in the steel compared to other beams. Main dapped end steel was stressed similarly in all cases.

5. Conclusions

A parametric study was conducted on reinforced dapped-end beam models with different reinforcement layouts. The following points can be highlighted from the parametric study:

1. Under impact load, shear strength of DEBs increases as the concrete compressive strength increases.
2. Reinforcement layouts and concrete compressive strength didn't affect the failure mode of dapped end beams under impact load, all DEBs failed in shear.
3. Under the impact load all dapped end beams failed in shear as different parameters varied.
4. Cracks started to initiate in the re-entrant corner as expected due to the concentration of stresses.
5. Hanger and main dapped end reinforcement detailing affect sufficiently in crack control but not enhancement of shear strength for dapped-end beams under impact load.
6. Amount of provided main dapped end steel reinforcement has a significant impact on diagonal shear cracks and damage of beam under impact load.
7. Main dapped end steel reinforcement showed more ductile behavior in case of higher concrete compressive strengths. Similarly, more ductile behavior was noticed on main dapped end reinforcement steel bars as the provided area of steel increased.

6. References

[1] Ahmad S, Elahi A, Hafeez J, Fawad M and Ahsan Z 2013 Evaluation of the Shear Strength of Dapped Ended Beam. *Life Science Journal,* **10**(3), p 1038

[2] Lu W Y, Lin I J, Hwang S J and Lin Y H 2003 Shear strength of high-strength concrete dapped-end beams. *Journal of the Chinese Institute of Engineers,* **26**(5), p 671

[3] Aswin M, Syed Z I, Wee T and Liew M S 2014 Prediction of Failure Loads of RC Dapped-End Beams *Applied Mechanics and Materials*, v **567**, p 463

[4] Wang Q, Guo Z and Hoogenboom P C 2015 Experimental investigation on the shear capacity of RC dapped end beams and design recommendations. *Structural Engineering and Mechanics, v* **21**, p221

[5] Aswin M, Bashar M, Liew M S and Syed Z I 2015 Shear failure of RC dapped-end beams. *Advances in Materials Science and Engineering*

[6] Desnerck P, Lees J M, and Morley C T 2016 Impact of the reinforcement layout on the load capacity of reinforced concrete half-joints. *Engineering Structures v* **27**, p 227

[7] Zhan T, Wang Z and Ning J 2015 Failure behaviors of reinforced concrete beams subjected to high impact loading. *Engineering Failure Analysis v* **56**, p 233

[8] Adhikary S D, Li B and Fujikake K 2012 Dynamic behavior of reinforced concrete beam s under varying rates of concentrated loading. *International Journal of Impact Engineering v* **47**, p 24

[9] Adhikary S D, Li B and Fujikake K 2014 Effects of High Loading Rate on Reinforced Concrete Beams". *ACI Structural Journal, v* **111**(3), p 651

[10] Saatci S 2007 Behaviour and modelling of reinforced concrete structures subjected to impact loads. *PhD dissertation* (University of Toronto, Toronto)

[11] Syed Z I, Hejah E S and Mohamed O A 2016 Modelling of Dapped-End Beams under Dynamic Loading. Int Journal of Mechanical Engg and Robotic Research, **6** (3), p 242

[12] Belytschko T, Liu W and Moran B 2000 *Nonlinear Finite Elements for Continua and Structures*, (Chichester: John Wiley & Sons Ltd)

[13] LSTC, LSDYNA Keyword's Manual 2016 Volume II Material Models (Livermore Software Technology Corporation, Livermore, California)

Identification and correlation analysis for performance shaping factors in flight crew operation

Lijing Wang[1*], Yanlong Wang[1], Yingchun Chen[2] and Dayong Dong[2]

[1] School of Aeronautic Science and Engineering, Beihang University, 37 Xueyuan Road, Beijing, China
[2] Shanghai Aircraft design and research Institute, Commercial Aircraft Corporation of China Ltd, 5188 Jinke Road, Shanghai, China

*wanglijing505@126.com, wyllaf@sina.com

Abstract. The dependency between PSFs (performance shaping factors) is gaining increasing attention in HRA (human reliability analysis). In this paper, 79 PSFs were identified through literature review and discussion of focus group, which is composed of human factors/HRA specialists and civil pilots. These PSFs were classified into 10 categories as cognitive characteristics, physiological and psychological characteristics, personal and social characteristics, procedures, task characteristics, human-machine interface, system state, phenomenology characteristics, physical working conditions team and organizational factors. Then a survey of 299 pilots was conducted. A self-rate scale was used to investigate how the pilots were influenced by these PSFs. Correlation analysis shows that the correlations between PSF subsets are moderate to strong. The result suggests further research on the dependency between PSFs and PSF interactions need to be included in future HRA efforts.

1. Introduction

The behaviour of human beings in complex systems is affected by many different factors, which may be external factors or some characteristics of human beings themselves. These factors are named performance shaping factors (PSF) [1]. PSFs can influence operator's performance and human error probability (HEP), which may be positive or negative. PSF is an important concept in human reliability analysis (HRA) first proposed by Swan and Guttmann in the THERP method, and has been used in most of the later HRA methods to estimate HEPs. In different HRA methods, PSF might be called different names, including performance influence factor(PIF), common performance condition(CPC) [2], error producing condition(EPC) [3]and error forcing context(EFC)[4].

The use of PSF in HRA can be summarized into the following categories:

- HEP quantification, such as SLIM [5], INTENT [6], STAHR [7] and HRMS [8];
- Analysis of error of commission, such as Macwan's PSF taxonomy [9], Julius's PSF taxonomy [10]and ATHEANA [4];
- Overall context assessment and error analysis, such as CREAM [2], HRMS and INCORECT [11];
- HRA database, such as Taylor-Adams's PSF taxonomy for CORE-DATA [12].

In general, the PSF classification is for specific purposes and application domains. Each HRA method has its own PSF sets. Currently, most HRA methods and PSF taxonomies are developed for the nuclear power sector. There are few PSF taxonomy for aviation transportation. A review of HRA empirical studies called for the identification of a key subset of PSFs [13,14]. The review also asked the question whether HRA methods using a key subset of PSFs and their corresponding qualitative analysis and quantification process can produce reliable and reasonable HEPs for most scenarios. Liu et al argued this vital question should be addressed in future HRA empirical studies [15]. Therefore, in order to ensure a consistent and effective human reliability analysis, it is necessary to establish a key PSF subset for pilots considering the task and context features in flight operation.

Moreover, dependency between PSFs is an important issue in HRA. There are two types of interrelationships, i.e., moderating effects and mediating effects [16]. Although, HRA specialists recommend taking into account for the dependencies and interactions, there is no agreement among methods on how to quantify the effects and interactions of PSFs within a task [17]. Many HRA methods assume PSFs are independent each other (such as SLIM [5], HEART [3], THERP [1]), while some other methods do not consider the correlations among PSFs explicitly (such as HERA)[18].

However, the dependency between PSFs is gaining increasing attention in HRA [19, 20]. This issue is related to the PSF causal model, which is required as an important input by several novel HRA methods, especially those based on Bayesian networks [21, 22]. The PSF dependency model in CREAM [2] was built based on human factors and HRA expert analysis and judgment. Another method for developing the PSF dependency model is gathering the opinions of domain experts and analyzing them through statistics. This is what employed in this paper.

The purpose of this paper is to identify the key subset of PSFs for civil flight crew and to evaluate the correlations among them through a large-scale pilot survey. The other parts of the paper are arranged as follows: The second section describes the process and results of identifying the key PSF subset for flight crew; the third section introduces the survey questionnaire and investigation process briefly; the fourth section provides the correlation analysis results and discussions; and the last section gives the conclusion.

2. Identification of flight crew PSFs

Many HRA methods have already been developed with considerations of various PSF sets. The inconsistence of PSFs selected may lead to some serious issues. First, the use of different PSFs makes the comparison between HEPs calculated by different methods meaningless. Moreover, to reduce human errors, corresponding measures should be taken according to the PSFs selected. The second problem is the number of PSFs used. Some methods use a limited number of PSFs, which can cause analysts to overlook important factors and underestimate the contribution of human error to the overall system safety. The third problem is that the definition and description of each PSF is distinct, which may lead to inconsistent evaluation of the same PSF by different evaluators, and thus different HRA results. Therefore, the above issues should be considered when identifying the key subset of PSFs for flight operation.

Kim and Jung's [23] review on PSF taxonomy forms the basis of this paper. In their article, they reviewed 18 PSF taxonomies and collected 220 PSFs. Kim and Jung [23] argued that the operator task context model can be demonstrated as Fig.1. According to figure 1, the collated PIFs are classified into four main groups: human, system, task, and environment. The boundary of each group is defined as follows.:

- Human: Personal characteristics and working capability of the operator;
- Task: Characteristics of the procedures and tasks need to be completed;
- System: Human-Machine Interface, plant hardware system and physical characteristics of plant process;
- Environment: Team and organization factors, and physical working conditions.

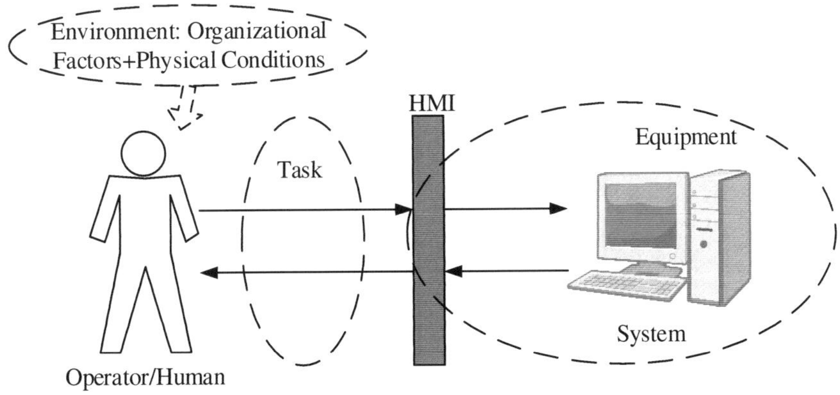

Figure 1. The context model of operator task.

The 220 PSFs collected by Kim and Jung [23] served as the complete set of PSFs for civil flight crew. Then a focus group composed of three human factors/HRA specialists and two pilots was set up to identify the key PSF subset. The below principles were followed during the screening process.

- The PSF selection should include all important factors in the assessed task context as much as possible;
- The selected PSF should not overlap with each other;
- Choose factors that directly affect the occurrence of human error;
- The selected PSF could be reflected in HRA;
- The selected PSF can be evaluated in practice;
- The terms describing PSFs should be as practical as possible and easy to be understood.

Then 79 key PSFs were identified for flight crew through focus group composed of three human factors/HRA specialists and two airline pilots. These PSFs were classified into 10 categories, i.e., cognition characteristics (CC), physiological and psychological characteristics (PPC), personal and social characteristics (PSC), procedures (P), task characteristics (TC), human machine interface (HMI), system state (SS), Phenomenological characteristics (PC), physical working conditions (PWC) and team and organization factors (TOF) (table 1).

Table 1. Key PSF subset for civil flight crew operation.

PSF Groups	PSF Categories	PSFs
Human	Cognition Characteristics(CC)	Attention
		Skill level
		Experience
		Operator Diagnosis
		Perceived Importance
		Confidence in Diagnosis
		Memory of Previous Actions and Accident History
	Physiological and Psychological Characteristics (PPC)	Fatigue
		Discomfort
		Emotion
		Confusion/Perplexity
		Task load
		Fear of Failure/Consequences
	Personal and Social Characteristics (PSC)	Attitude
		Motivation
		Risk Taking

		Self-confidence
		Sense of Responsibility
		Role/Responsibility
Task	Procedures (P)	Usability
		Quality
		Level of Detail
		Number of Steps
		Required Time for Completion
		Level of Standardization in Use of Terminology
		Decision Making Criterion
		Logic Structure
		Number of Simultaneous Tasks
		Adequacy of Caution/Warning
	Task Characteristics (TC)	Task Type:Procedure Following, Monitoring, Detection, Verification, Diagnosis, Recovery
		Required Level of Cognition
		Dynamic VS. Step-by-Step Activities
		Number of Required Information
		Number of Necessary Information to Be Memorized
		Information Load
		Task Difficulty
		Task Novelty
		Frequency and Familiarity of Task
		Number of Simultaneous Goals/Tasks
		Discrepancy between Training and Reality
		Perceptual Requirements
		Task Criticality
		Degree of Manual Operation
		Precision
		Requirement on and Type of Feedback
		Communication Requirement
		Team Cooperation Requirement
System	Human Machine Interface (HMI)	Availability
		Discrimination/Distinguishability of Signals
		Control–Display Relationships
		Existence of Failed Indicator
		Reachability
		Visibility
		Complicatedness of Control Panel
	System State (SS)	Inherent System Complexity
		Number of Coupled Components
		Level of Automation
	Phenomenological Characteristics (PC)	Number of Dynamic Changing Variables
		Time Available for Operator Performance
		Time Pressure
		Degree of Alarm Avalanche

Environment	Physical Working Conditions (PWC)	Temperature/Humidity/Pressure/Illumination
		Interference in Communication
		Noise
		Vibration
		Narrow Work Space or Obstacles
		Accessibility of Components
		Circadian Rhythm Effects
	Team and Organization Factors (TOF)	Clearness in Job Description or Role Definition
		Adequacy of Distributed Workload
		Intra/Inter-Team Cooperation
		Ability/Leadership/Authority of Team Leader
		Frequency and Training Time
		Work/Rest Schedule
		Shift Rotation
		Maintenance
		Rewards and Punishments
		Routine Violations
		Openness in Communication

3. Questionnaire and pilot investigation
A pilot self-rating scale was developed after identifying the critical PSF subset for the flight crew operation of civil aircraft. In pilots' daily mission, to what extent they were influenced by these PSFs were investigated using the five-degree scale. And the score 1 to 5 represent levels of very little, little, moderate, large and very large, respectively. In addition to these questions, flight hours, flight level, and aircraft types ever piloted were also asked. A total of 299 pilots participated in the survey, producing 231 valid questionnaires, with an effective rate of 77.3%. The flight hours is from less than 100 hours to 30,000 hours, with an average of 6,386 flight hours (SD=6921).

4. Result and discussion
To reduce the amount of analysis, the correlation analysis was only performed at a higher level of PSF subset, i.e. the second level including 10 PSF categories. Each PSF category of the second level was treated as a PSF in the correlation analysis. Each PSF score in the second level is the mean score of each PSF in its lower level. The descriptive statistics result is shown in table 2. The result shows that the influences of the 10 PSFs on pilot performance are between the level of "moderate" and "large". Among them, physiological and psychological characteristics, cognitive characteristics and team and organizational factors exert the greatest impact on pilot performance.

Table 2. Descriptive statistics

	N	Min	Max	Mean	SD
Flight Hours	231	100	30000	6835.92	6921.24
Cognition Characteristics	231	1.00	5.00	3.54	.69
Physiological and Psychological Characteristics	231	1.00	5.00	3.66	.74
Personal and Social Characteristics	231	1.00	5.00	3.29	.82
Procedures	231	1.00	5.00	3.40	.85
Task Characteristics	231	1.00	5.00	3.36	.75
HMI	231	1.00	5.00	3.22	.83
System State	231	1.00	5.00	3.26	.89

Phenomenological Characteristics	231	1.00	5.00	3.48	.87
Physical Working Conditions	231	1.00	5.00	3.23	.83
Team and Organization Factors	231	1.00	5.00	3.50	.78

There are three commonly used correlation coefficients, i.e. Pearson、Spearman and Kendall. Because the degree of influence of PSF on pilot operation performance is a kind of discrete grade data, the Kendall correlation coefficient is selected in this paper [24]. The correlation analysis results are shown in table 3. It can be seen from table 3 that PSFs of team and organizational factors, system states, physical working conditions and phenomenological characteristics are significantly correlated with pilot's flight hours at the level of 0.01 (bilateral), while cognitive characteristics, physiological and psychological characteristics are significantly correlated with pilot's flight hours at the level of 0.05 (bilateral). Taking cognitive characteristics as an example, cognitive characteristics mainly include 7 low-level PSFs, including attention, skill level, experience, diagnosis, perceived importance, diagnostic confidence and memory of previous actions and accident history. These PSFs will be enhanced with the increasing of flight hours, and pilot will have more experience, higher skill level and increased confidence in fault diagnosis. Pilots also tend to be more heavily influenced by this past knowledge. However, due to the extremely weak correlation coefficient, this tendency is not prominent.

Table 3. Results of correlation analysis

			Flight Hours	CC	PPC	PSC	P	TC	HMI	SS	PC	PWC	TOF
Kendall tau_b	Flight Hours	corrcoef	1.000	.107*	.093*	.024	.075	.066	.084	.142**	.125**	.128**	.148**
		Sig.[a]	.	.020	.044	.608	.102	.142	.068	.003	.007	.005	.001
	CC	corrcoef	.107*	1.000	.465**	.454**	.462**	.468**	.424**	.359**	.376**	.341**	.374**
		Sig.	.020	.	.000	.000	.000	.000	.000	.000	.000	.000	.000
	PPC	corrcoef	.093*	.465**	1.000	.448**	.443**	.441**	.399**	.384**	.427**	.432**	.443**
		Sig.	.044	.000	.	.000	.000	.000	.000	.000	.000	.000	.000
	PSC	corrcoef	.024	.454**	.448**	1.000	.556**	.512**	.486**	.439**	.386**	.412**	.418**
		Sig.	.608	.000	.000	.	.000	.000	.000	.000	.000	.000	.000
	P	corrcoef	.075	.462**	.443**	.556**	1.000	.654**	.521**	.517**	.496**	.464**	.446**
		Sig.	.102	.000	.000	.000	.	.000	.000	.000	.000	.000	.000
	TC	corrcoef	.066	.468**	.441**	.512**	.654**	1.000	.625**	.610**	.624**	.470**	.515**
		Sig.	.142	.000	.000	.000	.000	.	.000	.000	.000	.000	.000
	HMI	corrcoef	.084	.424**	.399**	.486**	.521**	.625**	1.000	.701**	.591**	.571**	.544**
		Sig.	.068	.000	.000	.000	.000	.000	.	.000	.000	.000	.000
	SS	corrcoef	.142**	.359**	.384**	.439**	.517**	.610**	.701**	1.000	.640**	.590**	.538**
		Sig.	.003	.000	.000	.000	.000	.000	.000	.	.000	.000	.000
	PC	corrcoef	.125**	.376**	.427**	.386**	.496**	.624**	.591**	.640**	1.000	.522**	.574**
		Sig.	.007	.000	.000	.000	.000	.000	.000	.000	.	.000	.000
	PWC	corrcoef	.128**	.341**	.432**	.412**	.464**	.470**	.571**	.590**	.522**	1.000	.559**
		Sig.	.005	.000	.000	.000	.000	.000	.000	.000	.000	.	.000
	TOF	corrcoef	.148**	.374**	.443**	.418**	.446**	.515**	.544**	.538**	.574**	.559**	1.000
		Sig.	.001	.000	.000	.000	.000	.000	.000	.000	.000	.000	.

*significant at the level of 0.05
**significant at the level of 0.01
[a] Sig.(bilateral)

According to the correlation analysis results, there are significant correlations at the level of 0.01 (bilateral) between all PSFs. This indicates that it is very necessary to consider the correlation between PSFs when evaluating the impact of PSF in the HRA. One of the most relevant PSFs to cognitive characteristics is physiological and psychological characteristics, possibly because cognition is essentially a spiritual activity. Both procedures and task characteristics are significantly relevant to personal characteristics. This can be reflected in the interview where pilots mentioned that if an operation procedure is illogical and there are many steps, the pilot is more likely to risk violation and not following the standard procedures. The most correlated PSF to procedure is task characteristics, since task characteristics such as complexity, novelty, and number of simultaneous tasks all will influence the logical structure, time required and number of steps of procedure. The strong correlations between procedures, task characteristics, HMI, system state and phenomenological characteristics also indicate that these PSFs should be comprehensively considered during the system design process in order to reduce task difficulty and demand. In particular, HMI and system state have the greatest correlation coefficient, reaching 0.7. This is because the complexity and automation level of the system largely determine the way the system is displayed and controlled, as well as the phenomenological features of the system. Physical working conditions and team/organizational factors are moderately correlated with HMI, system state, and phenomenological characteristics (0.4~0.6). This may be because the design of the system to a certain extent determines the working conditions, task assignment and roles of crew.

5. Conclusion

Most current HRA methods lack a sufficient consideration of the dependencies between PSFs. Many scholars in the HRA field have emphasized that PSF dependencies should be included in future HRA method development. In this paper, through literature review and focus group discussion, the key PSF subset for civil flight crew is identified based on the existing PSF taxonomies. The key subset contains 79 PSFs, classified into four group of human, task, system and environment, and further divided into 10 categories as cognitive characteristics, physiological and psychological characteristics, personal and social characteristics, procedures, task characteristics, human-machine interface, system state, phenomenology characteristics, physical working conditions team and organizational factors. Then, the influences of these PSFs on pilot performance were investigated through a pilot self-rating scale. A total of 299 pilots participated in the survey. The survey found that pilots generally believed the PSF had a "medium" to "large" impact on operational performance. Moreover, correlation analysis was performed between PSFs and between PSF and pilot flight hours. The results showed that the PSFs of team and organizational factors, system state, physical working conditions, phenomenological characteristics, cognitive characteristics, physiological and psychological characteristics had a significant but weak correlation with flight hours. While the PSF categories were generally moderately (0.4~0.6) to strongly (0.6~0.8) correlated. This also reflects the complex interaction between the internal elements of civil air transport system as a complex socio-technical system.

This paper is only a preliminary attempt to evaluate the correlations between flight crew PSFs. In the future, we will continue to expand the sample size for a more detailed analysis of dependencies between flight crew PSFs. Furthermore, the specific dependence type between PSFs will be studied through mediating effect analysis and moderating effect analysis. In addition, the aviation accident report data will also be analyzed to explore the PSFs dependencies, providing inputs for the establishment of PSF interaction model, such as Bayesian Network, through the fusion of multi-source evidences.

Acknowledgments

The authors wish to acknowledge pilots from China Southern Airlines for their assistance in focus group discussion and survey data collection.

References

[1] Swan A D and Guttmann H 1983 *Handbook of human reliability analysis with emphasis on nuclear plant* (Washington DC: US Nuclear Regulatory Commission)

[2] Hollnagel E 1998 *Cognitive Reliability and error analysis method-CREAM* (Oxford: Elsevier Science)

[3] Williams J 1986 *9th advances in reliability technology Symp.* (University of Bradford) pp. B3/R/1 – B3/R/13
[4] Office of Nuclear Regulatory Research 2000 *Technical basis and implementation guidelines for a technique for human event analysis* (Washington DC: US Nuclear Regulatory Commission)
[5] Embrey D E, Humphreys P, Rosa E A, Kirwan B and Rea K 1984 *SLIM-MAUD: An approach to assessing human error probabilities using structured expert judgment* a *technique for human event analysis* (Washington DC: US Nuclear Regulatory Commission)
[6] Gertman D I, Blackman H S, Haney L N, Seidler K S and Hahn H A 1992 INTENT: a method for estimating human error probabilities for decision based errors *Reliab. Eng. Syst. Saf.* **35** 127-36
[7] Phillips L D, Humphreys P, Embrey D and Selby D L 1990 *Influence diagrams, belief nets and decision analysis* ed R M Oliver and J Q Smith (John Wiley & Sons) chapter 9
[8] Kirwan B 1997 The development of a nuclear chemical plant human reliability management approach: HRMS and JHEDI *Reliab. Eng. Syst. Saf.* **56** 107-33
[9] Macwan A 1994 A methodology for modelling operator errors of commission in probabilistic risk assessment *Reliab. Eng. Syst. Saf.* **45** 139-57
[10] Julius J, Jorgenson E, Parry G W and Mosleh A M 1995 A procedure for the analysis of errors of commission in a Probabilistic Safety Assessment of a nuclear power plant at full power *Reliab. Eng. Syst. Saf.* **50** 189-201
[11] Kontogiannis T 1997 A framework for the analysis of cognitive reliability in complex system: a recovery centred approach *Reliab. Eng. Syst. Saf.* **58 233**-48
[12] Kirwan B, Basra G and Taylor-Adams S E 1997 *Proc. of the 1997 IEEE 6th Conf. on Human Factors and Power Plants* vol 9 (Orlando: IEEE) pp7-12
[13] Forester J, Dang VN, Bye A, Lois E, Massaiu S and Broberg H. 2014 *The Int. HRA Empirical Study: lessons learned from comparing HRA methods predictions to HAMMLAB simulator data* (Washington, D C: US Nuclear Regulatory Commission)
[14] Forester J, Liao H, Dang VN, Bye A, Lois E and Presley M. 2016 *The US HRA Empirical Study: assessment of HRA method predictions against operating crew performance on a US nuclear power plant simulator* (Washington, D C: US Nuclear Regulatory Commission)
[15] Liu P, Lyu X, Qiu Y P, He J D, Tong J J, Zhao J and Li Z Z 2017 Identifying key performance shaping factors in digital main control rooms of nuclear power plants: A risk-based approach *Reliab. Eng. Syst. Saf.* **167** 264-75
[16] Liu P and Li Z Z 2014 Human error data collection and comparison with predictions by SPAR-H *Risk Anal* **34** 1706-19
[17] Hallbert B, Gertman D, Lois E, Marble J, Blackman H and Byers J 2004 The use of empirical data sources in HRA *Reliab. Eng. Syst. Saf.* **83** 139-43
[18] Leva M C, Ambroggi M D, Grippa D, Garis R D, Trucco P and Sträter O 2009 Quantitative analysis of atm safety issues using retrospective accident data: the dynamic risk modelling project *Saf. Sci.* **47** 250-64
[19] Zwirglmaier K, Straub D and Groth K M 2017 Capturing cognitive causal paths in human reliability analysis with bayesian network models *Reliab. Eng. Syst. Saf.* **158** 117-29
[20] Mkrtchyan L, Podofillini L and Dang V N 2016 Methods for building conditional probability tables of bayesian belief networks from limited judgment: an evaluation for human reliability application *Reliab. Eng. Syst. Saf.* **151** 93-112
[21] Musharraf M, Hassan J, Khan F, Veitch B, Mackinnon S and Imtiaz S 2013 Human reliability assessment during offshore emergency conditions *Saf. Sci.* **59** 19-27
[22] Musharraf M, Smith J, Khan F, Veitch B and Mackinnon S 2018 Incorporating individual differences in human reliability analysis: an extension to the virtual experimental technique *Saf. Sci.* **107** 216-23
[23] Kim J W and Jung W 2003 A taxonomy of performance influencing factors for human reliability analysis of emergency tasks *J. Loss Prev. Process Ind.* **16** 479-95
[24] Xue W 2004 *SPSS: statistics analysis methods and applications* (Beijing: Electronic Industry Press)

Structural reliability and performance evaluation of symmetric and asymmetric multi-storey buildings subjected to lateral wind forces

M A Kewalramani[1], Z I Syed[1], M S Almustafa[2], and M S Hassouna[3]

[1]College of Engineering, Department of Civil Engineering, Abu Dhabi University.
[2]Ali & Sons Contracting Company, Abu Dhabi, United Arab Emirates.
[3]Al Ittimad Foundations L.L.C, P O Box 2103, Abu Dhabi, United Arab Emirates.

Email: manish.kewalramani@adu.ac.ae

Abstract. Lately there has been enormous enhancement of buildings architecturally. Most of the buildings are designed and constructed with diverse variations like soft–storey, torsional irregularity, asymmetrical layout of walls, irregularities in plan and elevation and thus overall building configuration. Any such variation result in challenging implications in terms of structural reliability and appropriate design of structural members. Calculation of lateral forces on these structures is an essential process in case of high– rise structures and structural performance of these aesthetically appealing asymmetric irregular shaped buildings especially when subjected to heavy wind forces need to be carefully assessed by structural engineers. The present work intends at performance evaluation by studying structural reliability of symmetrical and asymmetrical regular square, L, C and T–shaped reinforced concrete multi–storey buildings subjected to wind loads. The wind load analysis on these buildings is carried out in line with ASCE 7–10 design code. A comparative evaluation of critical parameters like fundamental period, natural and circular frequency of building, maximum joint displacement, storey drift ratio, inter–storey drift, internal forces in structural members, storey response and storey stiffness determines that difference in shape of a particular storey (or building) influence overall structural performance of building.

1. Introduction

A building is usually considered as asymmetric in plan or in elevation based on the distribution of its mass and stiffness along each storey throughout the height of the building. Asymmetric nature of building may cause interruption of force flow and stress concentration when especially subjected to lateral loads. Buildings in general are subjected to different type of loadings such as superimposed dead loads, live loads, and lateral loads etc. Lateral loads are typically of two types i.e., wind loads and seismic loads. This study emphasize on wind loads since they are one of the most critical loads in tall structures like multi–storey buildings and it must be accounted when designing any such structure. Present study conducts a comparative structural reliability and performance evaluation of symmetrical and asymmetrical buildings subjected to wind loads using ETABS. The study compares significant parameters such as fundamental period, maximum deflection, story drifts, story stiffness, story response and storey internal forces obtained by modelling structural members of buildings with different plan configurations such as regular square shape, L-shape, C-shape and T-shape.

IOP Conf. Series: Materials Science and Engineering **575** (2019) 012005 doi:10.1088/1757-899X/575/1/012005

Fundamental period of a building is defined as amount of time a displaced point takes to get to its original position due to lateral force. Maximum deflection is the farthest displacement any point on the structure can reach due to different loading types, and in this case it is due to wind loading. Story drift is the lateral displacement of a floor level related to the floor above or below. Story response is a graph to represent the relationship between the story displacement and the story height of each building. As for the story stiffness, it is the story shear per story displacement. All such wind load and relevant parameters calculations for these four different shapes of buildings were done in accordance with ASCE 7–10 wind load provisions.

2. Methodology

The present work can be considered as a comparative and quantitative structural reliability and performance evaluation of four buildings with different plan configurations when subjected to wind loads. The analysis was conducted on symmetrical and asymmetrical reinforced concrete buildings with uniform panel dimensions of 5m×5m in order to find more suitable asymmetrical structure for resisting the wind loads efficiently. These four plan configurations of buildings consists of 15 typical stories, with typical story height of 3.5m and base story height of 4.5m. All the structural members including shear walls dimensions and locations are modelled identical to have a through comparison and collection of numerical data for parameters such as fundamental period, natural and circular frequency of building, other general parameters like maximum joint displacement, storey drift ratio, inter–storey drift, internal forces in structural members, storey response and storey stiffness affected due to wind loads as specified by ASCE 7–10. The following figures 1 (a), (b), (c) and (d) show plan configurations of buildings modelled using ETABS.

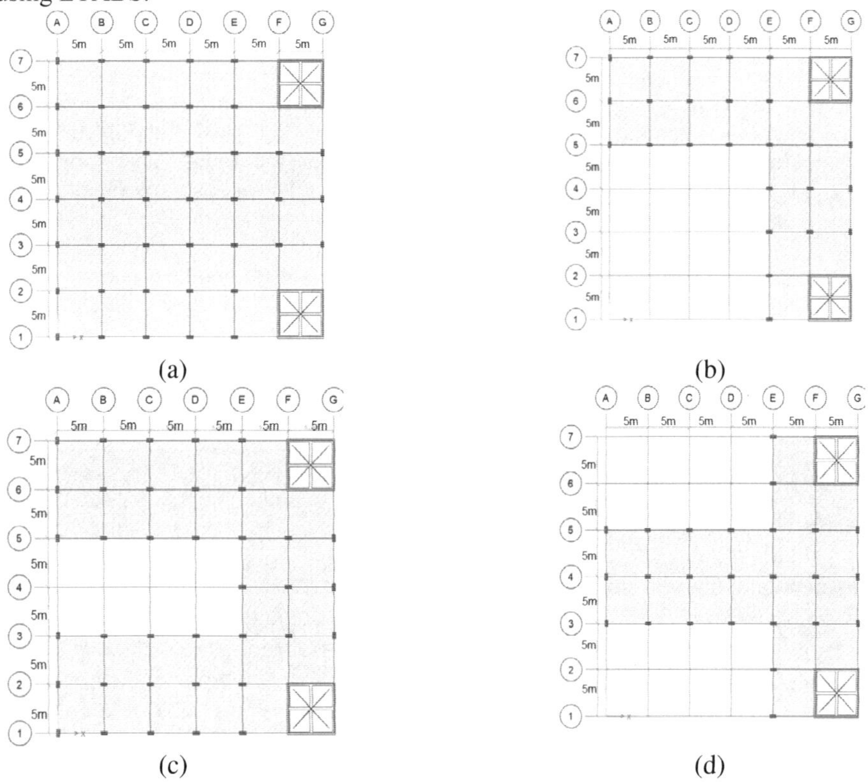

Figure 1. Building Plans (a) Regular Symmetrical Square (b) Asymmetrical L (c) Asymmetrical C (d) Asymmetrical T–Shape Buildings.

The purpose of analysis was to collect relevant data for structural reliability and performance evaluation of these four building models when subjected to wind loading case 1 and case 2 shown in figure 2 with a prevailing average wind speed of 100 mph in Abu Dhabi and a structure exposure class of B category using ETABS.

IOP Conf. Series: Materials Science and Engineering **575** (2019) 012005 doi:10.1088/1757-899X/575/1/012005

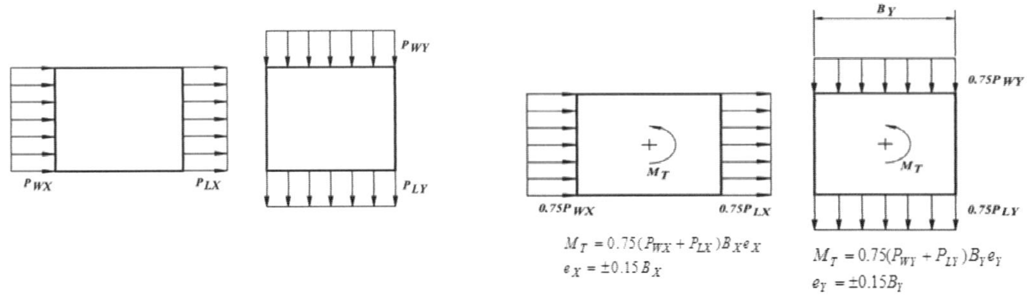

Case I Case II

Figure 2. Wind load cases considered in present analysis [1].

3. Results and Discussion

A comparative analysis of lateral load specific parameters like fundamental period, natural and circular frequency of building and some other general parameters like maximum joint displacement, storey drift ratio, inter–storey drift, internal forces in structural members, storey response and storey stiffness due to change in plan configuration of a building is carried out and discussed below.

3.1 Fundamental Period

Modal analysis of buildings is carried out to measure and analyze dynamic response of buildings when subjected to wind loads. Generally, the first mode of vibration is of primary interest as it has the largest contribution to the structure's motion. The natural frequencies of the structure are frequencies at which the structures naturally tend to vibrate when subjected to some kind of disturbance i.e. wind loads in present case. Every building has its own frequency and it is dependent on stiffness and mass of the building. A building with greater mass will have lower natural frequency [2]. A square shape building with higher plan area is found to have lower values of natural frequency, circular frequency and eigen value compared to other plan configurations as seen in Table 1.

Table 1. Modal periods and frequencies of buildings with different plan configuration

Parameters	Square Shape		L–Shape		C–Shape		T–Shape	
	Mode 1	Mode 2	Mode 1	Mode 2	Mode 1	Mode 2	Mode 1	Mode 2
Period (sec)	1.4	1.159	1.241	1.088	1.368	1.099	1.235	1.093
Freq. (cyc/sec)	0.714	0.863	0.806	0.92	0.731	0.91	0.809	0.915
Cir. Freq. (rad/sec)	4.4878	5.4228	5.0638	5.7776	4.5921	5.7172	5.0861	5.7504
Eigen Value (rad^2/sec^2)	20.1403	29.407	25.6424	33.3806	21.0869	32.6863	25.8688	33.0676

3.2 Joint Displacement due to Wind Loads

In general joint displacement at the top of stories of asymmetric plan configuration buildings is observed to be more in all the three directions compared to a square shape building. Joint displacements of C and T–shape buildings are however less than L–shape building as it is not symmetric about both *X* and *Y* axes. At the top story of building, a minimum joint displacement of 6.77×10^{-12} mm is observed for a C–shape building for wind load case 1 and a maximum joint displacement of 17.7mm is observed for a L–shape building for wind load case 2 which can be clearly seen in following figure 3 (a) and (b).

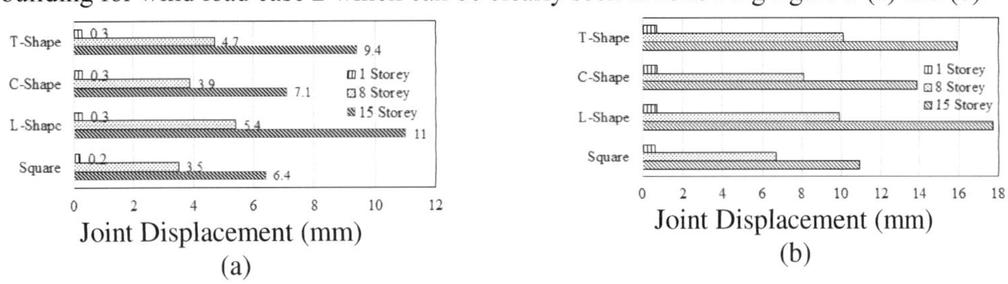

3.3 Story Drift

Storey drift or inter–storey drift is one of the particularly useful engineering response quantity and indicator of structural performance, especially for high-rise buildings when subjected to lateral loads [3]. Lateral deflection and drift have three primary effects on a structure; the movement can affect the structural elements (such as beams and columns); the movements can affect non–structural elements (such as the windows and cladding); and the movements can affect adjacent structures. Table 2 summarizes story drift (mm) and story drifts ratio of the first, eight and top storey relative to the base due to both cases of wind loads for various plan configurations.

Table 2. Storey Drift Ratio and Drift (mm) for various plan configurations due to two cases of wind loads

Storey	Max. Drift	Square Shape		L–Shape		C–Shape		T–Shape	
		Wind 2	Wind 1	Wind 2	Wind 1	Wind 2	Wind 1	Wind 2	Wind 1
15	*Max. Drift Y*	0.000132		0.000255		0.00018		0.000255	
	Max. Drift X		0.000097		0.000209		0.000112		0.000171
	X	29.2	25	0	25	0	25	0	25
	Y	25	2.5	30	0	30	27.5	20	2.5
8	*Max. Drift Y*	0.000246		0.000404		0.000315		0.000409	
	Max. Drift X		0.000148		0.000248		0.000164		0.000213
	X	0	27.5	0	29.2	0	27.5	0	27.5
	Y	30	2.5	30	0	30	27.5	20	2.5
1	*Max. Drift Y*	0.000137		0.000156		0.000148		0.000161	
	Max. Drift X		0.000055		0.000066		5.80E-05		6.30E-05
	X	0	29.2	0	29.2	0	29.2	0	29.2
	Y	30	0	30	0	30	30	20	0

(a) (b)

Figure 4. Maximum storey drift ratio (a) in X-direction and (b) in Y-direction

Figure 4 (a) and (b) above show maximum storey drift ratio in X and Y direction at storey 15, storey 8 and storey 1 of different plan configuration buildings. It can be clearly seen that a square shape building has minimum storey drift in both directions for all the three stories considered followed by C-shape and T-shape buildings.

3.4 Internal forces in structural members

Maximum and minimum values of internal forces like shear force, bending and torsional moments, deflection for a particular beam and axial force and torsional moment for a particular column most affected by the shape of building were considered for analysis as shown in Table 3 (for beam B67 in 8th Storey) and Table 4 (for column C5 in 8th Storey).

Table 3. Internal forces in Beam B67 in 8th Story due to Wind Loads

Internal forces	Square		C-Shape		L-Shape		T Shape	
	Max	Min.	Max	Min.	Max	Min.	Max	Min.
Shear (kN)	8.35	-0.125	8.31	-0.483	10.922	-1.22	10.94	1.54
Bending Moment (kN-m)	12.06	-13.45	14.61	-15.44	19.35	-20.35	19.32	-20.3
Torsional Moment (kN-m)	4.75	-3.68	4.29	-3.176	5.545	-4.537	5.55	-4.59
Deflection (mm)	-0.2	8.56×10^{-3}	-0.5	0.2	-0.6	0.3	-0.6	0.5

Location of these values within the member vary in some cases, however in most of the cases the location of maximum and minimum value remains same due to the similar nature of load considered for this part of structural reliability analysis. It can be seen from the table 3 and table 4 that the internal forces and deflection are more in case of asymmetrical shapes (L shaped building, in particular).

Table 4. Internal forces in Column C5 in 8th Story due to Wind Loads

Internal forces	Square		C-Shape		L-Shape		T Shape	
	Max	Min.	Max	Max	Min.	Min.	Max	Min.
Axial Force (kN)	74.09	-7.95	98.68	-81.85	114.29	-52.5575	152.51	-40.99
Torsion (kN-m)	0.443	-0.443	0.563	-0.563	0.914	-0.782	0.8703	-0.872

3.5 Story Response

Storey response is analyzed in terms of maximum storey displacement as shown in Figure 5. Storey displacement is the lateral displacement of the storey relative to the base. There is a maximum permissible limit of storey displacement specified in design codes [4]. It can be seen that a fifteen storey square shape building has lesser value (6.4 mm) of storey displacement compared to C and T – shaped building, while L shape building showed maximum storey displacement of 11.9 mm at the top of the building when subjected to wind loads.

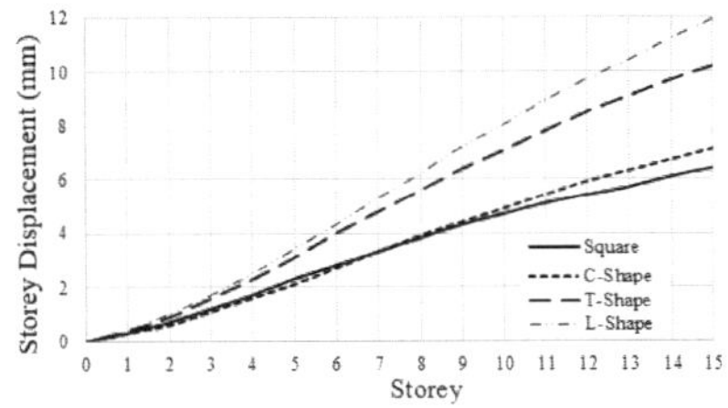

Figure 5. Storey Displacement (mm) measured for fifteen stories of different shaped buildings

Storey response is also studied in terms of minimum displacement in X and Y direction at all stories of different shaped buildings as shown in figure 6 (a) and (b). A similar behavior of displacement can be observed in both X and Y directions when wind loads are applied.

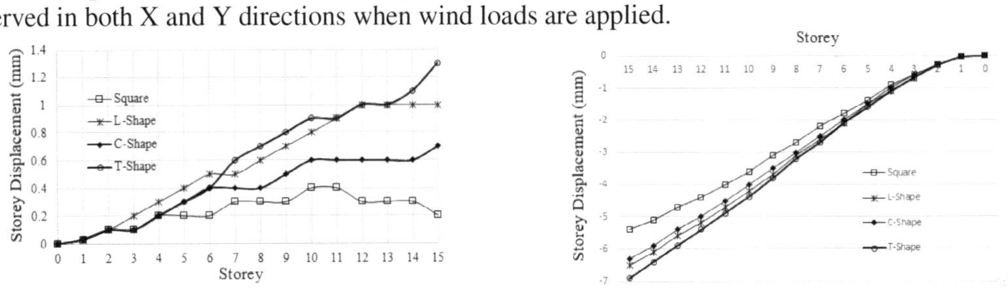

(a) (b)

Figure 6. Minimum storey displacements in (a) X- direction and (b) Y-direction

3.6 Storey Stiffness

The story stiffness is defined as inverse of inter–story drift when a unit load is applied at that particular story, in other words it is the shear force required to produce unit displacement of that storey. It is an important parameter for both static and dynamic analysis of building subjected to lateral loads, for calculation of lateral displacements in static analysis and calculation of lateral displacements and dynamic properties such as modal frequencies and shapes in dynamic analysis [5].

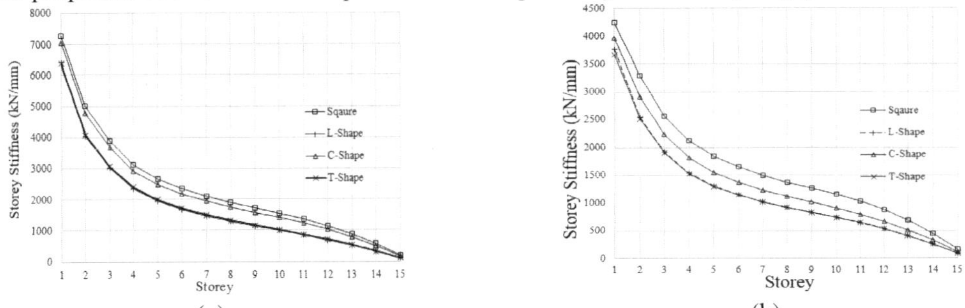

(a) (b)

Figure 7. Storey Stiffness (kN/mm) in (a) X-direction for wind load case I and (b) Y-direction for wind load case II

The inter–story stiffness is used as one the main damage–sensitive features of a multi-story building for detection, localization and quantification of damage caused. The distribution of lateral loads is dependent on the relative stiffness of each structure. Elements with higher stiffness resist more lateral load than elements with lower stiffness. The results of storey stiffness in X and Y direction are shown in figure 7 (a) and (b). A square shaped building has higher storey stiffness followed by C shape in both directions. However, for L and T-shaped buildings the results were quite similar as seen in Fig 7 (a) and (b).

4. Conclusions

A thorough comparison of parameters namely fundamental period, natural frequency and circular frequency, joint displacements, storey drift, internal forces in structural members, storey response in terms of maximum storey displacement, and storey stiffness of four different shaped buildings using ETABS has been carried out. From lateral load analysis of these buildings modal deflected shapes concluded that asymmetrical plans undergo more deformation than symmetrical plans.

Storey drift displacement increased with storey height up to 8th storey reaching to maximum value and then starts decreasing for wind load in X-direction while for wind load in Y-direction it increases from base till top of the building for all shapes. Relative to the square plan configuration, it can be observed that the L-shaped and T-shaped buildings have almost 48.24% increase in drift ratio in the Y-direction for storey 15, while the L-shaped have a 53.59% increase in drift ratio in the X-direction for storey 15. Fundamental period showed relatively significant decrease in the T-shaped building of almost 11.79% compared to the square shaped building which had the highest value of 1.4 seconds. Similarly, the L-shaped building had 11.35% lower fundamental period compared to square shape. Thus, it can be said that buildings with lower mass tend to have lower fundamental period or higher frequency.

It was also noticed that the storey stiffness of these plan configurations increase as the mass of the building increase since the square-shaped building showed roughly about 34% increased stiffness in the Y-direction when compared to L-shaped and T-shaped buildings, the same storey stiffness increase in X-direction was observed to be 30%.

In general it can be concluded that plan configurations of a building has significant impact on the lateral load response of structure in terms of displacement, story drift and story shear. Large displacements were observed in L–shaped building when subjected to wind loads in both X and Y-direction due to severe irregularity in plan i.e. asymmetric about both the axes. Hence, it can be said that T–Shape and L–shape multi-storeyed buildings are more susceptible to wind loads compared to the symmetrical buildings based on criteria included in present study.

References

[1] Bhandari N M, Krishna P and Kumar K 1987 *An Explanatory Hand book on Wind Loads on Building and Structure* Indore IS 875 (Part 3)

[2] Mendis P, Samali B and Cheung J 2007 *Wind Loading on Tall Buildings* EJSE Special Issue - Loading on Structures 41-54

[3] Thambiratnam D P and Thevendran V 1992 *Simplified Analysis of Asymmetric Buildings Subjected to lateral Loads* Elsevier - Computers & Structure **43**(5) 873-880

[4] Guleria A 2014 *Structural Analysis of a Multi-Storeyed Building using ETABS for different Plan Configurations* International Journal of Engineering Research and Technology, **3**(5) 1481-1485

[5] Mahesh S and Rao B P 2014 *Comparison of analysis and design of regular and irregular configuration of multi-story building in various seismic zones and various types of soils using ETABS and STAAD* IOSR Journal of Mechanical and Civil Engineering **11**(6) 45-52

[6] *Minimum Design Loads for Buildings and Other Structures* 2010 American Society for Civil Engineers (ASCE 7-10)

IOP Conf. Series: Materials Science and Engineering **575** (2019) 012006 doi:10.1088/1757-899X/575/1/012006

Runtime Contracts Checker: Increasing Robustness of Component-Based Software Systems

M Illarramendi, L Etxeberria, X Elkorobarrutia and G Sagardui

Mondragon Goi Eskola Politeknikoa S.Coop, Loramendi 4, Mondragon, Spain

millarramendi@mondragon.edu

Abstract. Software Systems are becoming increasingly complex leading to new Validation &Verification challenges. Model checking and testing techniques are used at development time while runtime verification aims to verify that a system satisfies a given property at runtime. This second technique complements the first one. This paper presents a runtime contract checker (RCC) which checks a component-based software system's contracts defined at design phase. We address embedded systems whose software components are designed by Unified Modelling Language-State Machines (UML-SM) and their internal information can be observable in terms of model elements at runtime. Our previous research work, CRESCO (C++ REflective State-Machines based observable software COmponents) framework, generates software components that provide this observability. The checker uses software components' internal status information to check system level safety contracts. The checker detects when a system contract is violated and starts a safeStop process to prevent the hazardous scenario. Thus, the robustness of the system is increased.

1. Introduction

The scope, complexity, and pervasiveness of software systems continue to increase dramatically. The consequences of these systems failing can range from mildly annoying to catastrophic. Software increasingly assumes the responsibility for providing functionality in systems. Therefore, improving robustness through software has a significant impact on the system.

Verification and validation techniques applied during development give a certain level of confidence in the correctness and are effective at detecting and avoiding anticipated faulty scenarios. However, in embedded software systems where a fault can lead to a critical scenario, we need a way to detect and mitigate hazardous and uncertain scenarios. Runtime verification techniques could be used to maintain safe control in unanticipated circumstances.

Monitoring information related to the internal status of the embedded software can anticipate the detection of a malfunction. This makes it possible to take corrective actions earlier and prevent faulty scenarios. This idea is described as a safety bag in [1] and [2]. The goal is to prevent software systems' hazardous states by means of safety verification at runtime. Thus, we increase their robustness and reliability.

Current runtime checking solutions as shown in Figure 1, are specified at different abstraction levels: system, component, class, method or statement. In most of the approaches, the checking is performed at the same level, that is, only by checking system level properties could system misbehaviour be detected. Nevertheless, component or class level properties can give valuable information in detecting system level problems and undesired emergent behaviour.

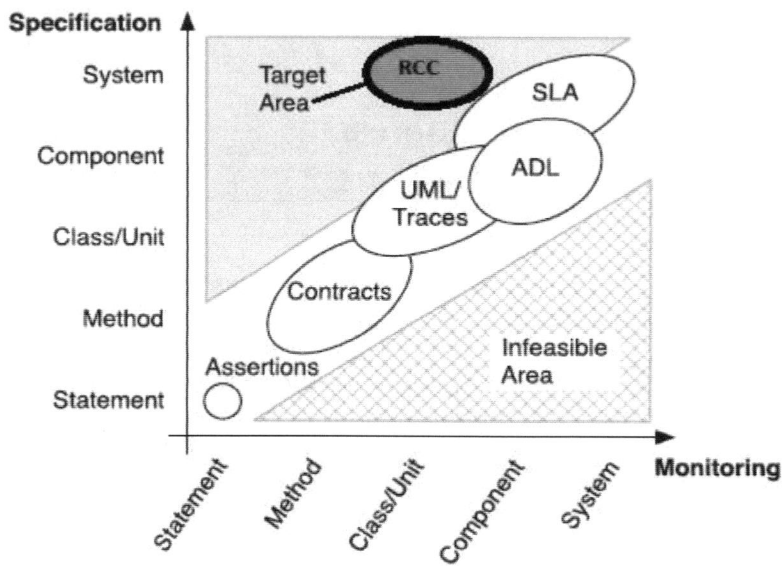

Figure 1. Abstraction level of specifications vs. runtime monitoring abstraction levels based on [3].

Most runtime software checkers require modifying the source code of the observed system by instrumented code. However, it is desirable that runtime checkers for testing safety properties of the systems should be isolated from the target system to minimize any disruption of the system being tested [4].

In this paper, the Runtime Contracts Checker (RCC) is presented. RCC tries to solve the aforementioned limitations of the current software runtime checkers. The checker is based on the concept of runtime Safety Contracts (SC).

Our approach (RCC) is classified in the target area of Figure 1. Specification is made using system level safety contracts and the monitored information is the internal status of the software components of the system. It verifies the system level safety contracts based on the internal status information received from each of the systems' software components' observable states at runtime. These contracts are checked when the internal status of the system components is going to perform a state transition. In this sense, our software checker shares the advantages of hardware checkers because we can detect the fault before a change in components' state happens. RCC is developed in an isolated way and therefore there is no need to change any code, nor the design of the system.

Main contributions of RCC are:

- check system level safety contracts by monitoring internal status of the system's software components,
- prevent errors before a change in a state of software components occurs,
- isolate the system's functionality and its own (RCC) while not interfering with the developer's design and development work.

It is worth noting that the RCC has been developed taking into account a previous work called CRESCO framework [5]. This framework is able to generate reflective software components that provide information about the internal status of the software components in terms of UML-SM elements at runtime. The contracts that the RCC checks are defined using the information provided by these software components at runtime. Nonetheless, the RCC solution can be used independently of CRESCO software components. In any case, the software components we are addressing have to fulfil the following conditions: (1) they have to be designed by UML-SMs and (2) they have to provide the internal status of their observed states at runtime.

RCC also requires consistent snapshots of the system and, to this end, for example, the observed system has to be a synchronous one or the system's messages must be causally ordered.
Section II introduces the Runtime Contract Checker. In Section III related work is detailed and, finally, section IV ends the paper with our conclusions and future lines of action.

2. Runtime contract checker

Component-based software systems are composed of various software components that interact to provide a given system functionality. The aim of the present work is to generate a checker to avoid errors and hazardous scenarios of component-based software systems at runtime. To this end, system level runtime safety contracts, based on internal status information of the software components, are defined and the specific checker is generated automatically.
This section presents the process of generating the RCC, the architecture of the contracts checking system, the internal status information to be checked by the RCC and how the runtime state-based safety contracts are specified.

2.1. Process for defining Safety Contracts and generating the RCC

The process to generate the RCC is embedded in a typical design process for developing dependable systems. After performing software system design phase and obtaining the system architecture with the decomposition of software components, together with a first design of the software components including their behaviour (UML-SM diagrams), the process for defining state-based safety contracts starts. This process has four steps (see Figure 2):

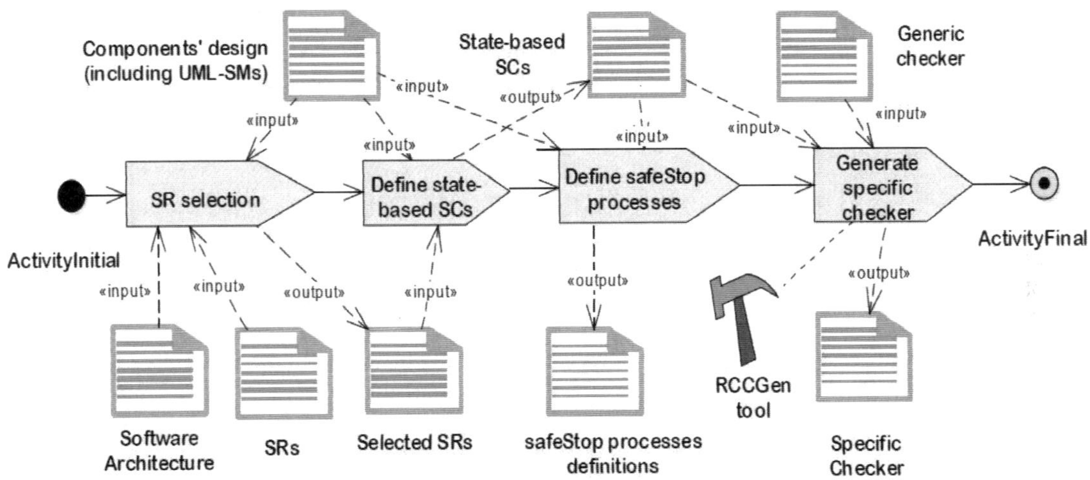

Figure 2. Process for defining state-based safety contracts.

- Select safety requirements (SR) to be used at runtime verification. Not all safety requirements are verifiable in terms of internal states of system's components; those that can be verified in this way have to be selected. The result is a list of safety requirements (SR_i).
- Define state-based safety contracts (SC) based on selected safety requirements.
- Define safeStop processes to be launched in case safety contracts are not fulfilled at runtime (a process for each safety contract).
- Generate the checker: RCCGen tool transforms the safety contracts to RCC Code (checker, in C++) automatically. RCCGen uses a generic checker as a basis and adds the specific safety contracts to the checker.

After these steps, the development continues with the implementation, verification and validation and integration steps.

2.2. RCC Architecture

In Figure 3, the overview of component-based software system's safety contract checking architecture is shown. The software components of the system are modelled by UML-SM and the runtime contract checker has two main components: the runtime contracts checker (RCC checker) and the safeStop processes manager.

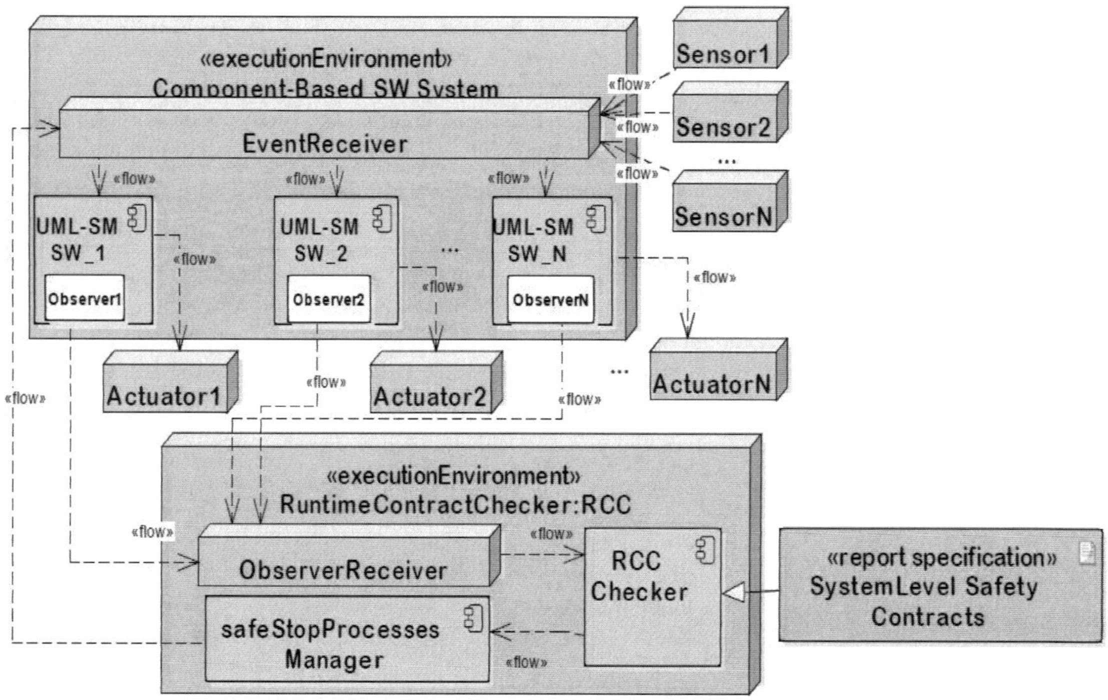

Figure 3 General Architecture of the Contracts Checking System.

RCC behaviour at runtime: The RCC starts after getting an initial consistent snapshot of the component-based software system. Next, the checker waits until it receives an update from any of the system's software components. The observers of these software components send their internal status information before performing a state transition. The RCC checker compares this information with the system level contracts. If system contracts are fulfilled, the system status information is updated and it waits for new updates.

In the event of the contracts not being fulfilled, the RCC checker notifies that circumstance to the safeStop processes manager. This manager starts the predefined safeStop process. This process sends safeStop events to the software components that are involved in the safeStop process. The process is dependent on the system and it is not in the scope of our work.

2.3. Internal status information of the monitored software components

This work addresses software components that are modelled by UML-SM and systems that are composed of several different of these components. Many tools can generate UML-SM based software components, one of which is the CRESCO framework. In addition, CRESCO creates software components that can reflect at runtime the UML-SM model from which they come.

IOP Conf. Series: Materials Science and Engineering **575** (2019) 012006 doi:10.1088/1757-899X/575/1/012006

For runtime checking, some decisions must be taken about what to observe. The RCC needs to receive runtime information from the software components of the system to be checked, in particular from the states annotated as observable.

2.4. Specification using Safety Contracts

We need to prove that the composite implementation of the system guarantees system level contracts at runtime. In our case, a safety contract specifies properties related to the internal behavior of the software components that are part of the system in terms of their UML-SM model (active states). That is what we call a state-based safety contract.

In our approach a system (Sys) may be composed of subsystems (that could be further decomposed) and primitive components (C) that cannot be further decomposed. Furthermore, the primitive components have a behavior specified using a state machine.

Safety requirement will be allocated to the system and a safety requirement may be satisfied by a safety contract or a set of safety contracts. A safety contract will be related to the states of the components involved in the contract. A grammar with regular expressions is used to specify what to check.

contract	:=	*constraint* | *timedConstraint;*
constraint	:=	*condition* **implies** *condition;*
condition	:=	*activeState* | **not** *condition* | *(condition)* | *condition* **or** *condition* | *condition* **and** *condition*;
timedConstraint	:=	*constraint* **in** *timeUnits*;
activeState	:=	*ComponentName:StateName;*

With this grammar, it is possible to specify constraints regarding the active states of components of the system. For instance, we can specify that the active state of the C_ {Door} must be S_ {Closed} when the active state of the C_ {Traction} is S_ {On}.

CTraction:SOn **implies** *CDoor:SClosed*

3. Related work

Creating a runtime checker for checking system level properties is an important research problem. Existing approaches to runtime verification require specifications that not only define the property to monitor, but also contain details of the implementation, sometimes even requiring the implementation to add special variables or methods for monitoring.

In [6], they defined a generic software monitoring model and analyzed different existing monitors. Depending on the programming language used and the formalism used to express the properties, different implementations of monitors have been proposed, among others CoMA, RV-MONITOR and AspectC++, BEE++, DN-Rover, HiFi etc. The abstraction level of specification and monitoring of the above solutions is the same: specifications and the monitor's checking properties are at the component or class level.

LuMiNous [3] framework's target is the same area as the checker presented in this paper: system level specifications are checked by components' level information. The contribution of this framework relies on the translation of high-level specifications into runtime monitors but, in this case, their solution is for Java (AspectJ based solution), which is not suitable for embedded systems.

4. Conclusion and future work

This paper presents an RCC that is automatically generated and which considers system level specific safety contracts to be verified at runtime.

The main conclusions are that the checker is able to detect all the safety related faults at runtime, thereby increasing the robustness and reliability of the system. As it uses components' internal

information, it has the ability to prevent faulty scenarios before having changed the system output signals. This is a benefit compared with software monitors that can only check the output signals.
Our last conclusion is that the process to implement and generate the RCC is cost-effective as it is generated automatically. The safety engineer has to define just the safety contracts (following the defined grammar to this end) and the safeStop processes. The rest of the process is automatic. The software developer of the software components only considers the functional aspects of the system.
As future research lines we might consider the following topics:

- Expand the empirical evaluation by using realistic use cases in different industrial domains and projects.
- Provide a tool or mechanism that will assist in adding the specific part of the safeStop processes systematically to the safeStop process manager.

Acknowledgments
The project has been developed by the Embedded System Group of MGEP and supported by the Department of Education, Universities and Research of the Basque Government under the projects Ikerketa Taldeak (Grupo de Sistemas Embebidos) and TEKINTZE (Elkartek 2018) and the European H2020 research and innovation programme, ECSEL Joint Undertaking, and National Funding Authorities from 19 involved countries under the project Productive 4.0 with grant agreement no. GAP-737459 - 999978918.

References

[1] IEC 2010 *61508: Functional safety of electrical/electronic/programmable electronic*
[2] B. Schön, M. Brini and P. Crubillé 2017 *Complementary methods for designing safety necessities for a Safety-Bag component in experimental autonomous vehicles* in Proceedings 12th National Conference on Software and Hardware Architectures for Robots Control
[3] J. Wuttke and M. Pezzé 2016 *Model-driven generation of runtime checks for system properties* Int J Softw Tools Technol Transfer.
[4] P. Koopman, A. Kane and T. Fuhrman 2014 *Monitor Based Oracles for Cyber-Physical System Testing* in Dependable Systems and Networks: Practical Experience Report
[5] M. Illarramendi, L. Etxeberria, X. Elkorobarrutia and G. Sagardui 2017 *Increasing Dependability in Safety CPSs Using Reflective Statecharts* in Computer Safety, Reliability, and Security
[6] G. Chang-Guo, J. Zhu and X.-L. Li 2011 *A Generic Software Monitoring Model and Feature Analysis* Journal of Software, vol 6, no 3, pp 395-403

IOP Conf. Series: Materials Science and Engineering **575** (2019) 012007 doi:10.1088/1757-899X/575/1/012007

Medical Device Integrated System Accelerated Life Testing

Khaled Atua (PhD) & Saeid Baniasad (PhD)[1]

[1]System Engineering Services, Baxter International Inc., Round Lake, IL, USA

khaled_atua@hotmail.com; saeid_baniasad@baxter.com

Abstract. This paper defines the accelerated life test design and technique to evaluate medical device system reliability. The proposed method includes developing system functional diagram to capture the overall ecosystem of the device and all control and noise factors that affect device performance. The test simulates real life therapy level stresses including temperature, patient use, disposable setting, power source alterations, medical prescription and duty cycle. These operation elements are used at different rates to balance the aging of different subsystems and rate of different failure mechanisms. The sample size of the test units is determined based on minimizing the overestimation error of the Mean Time Between Failure (MTBF) value. All observed events, and failures are categorized, and identified as failure per the expected complaint handling systems in real life. The test/tester setup errors and customer errors can also be tracked separately in the test and a mechanism of feeding back to human factors analysis, manufacturing, and service departments are identified to help optimize and inform product design and sustaining engineering activities. Test results and findings are to be compared to field data and benchmarked against a predicted number of failures based on the reliability requirements analysis established prior to product release. Test data and performance parameters are to be tracked and used to establish and optimize proactive and predictive maintenance interval and strategies for limited life and degrading components. Finally, a standardized testing method is to be revised based on lessons learned, and to be generalized for future product tests.

1. Introduction

It is well-established that demonstrating medical device reliability at the system therapy level is challenging in many dimensions. Accelerated life testing under laboratory conditions is one technique often used to simulate the complex real life usage of the medical device in the field. However, the traditional approach of component/subsystem/system-level reliability testing & analysis often allows some system and user interaction failures to escape to the field. The methodology described in this paper attempts to address this issue by establishing a dynamic profile of real-life therapy conditions to expose system interactions and failure modes by managing the signal and control factors, and by including many ecosystem noise factors under controlled laboratory conditions. The methodology also addresses the rationales for identifying the proper sample size for system testing that minimizes the error of overestimating the reliability (Mean Time Between Failure) from in-house lab testing before product is released to the field.

Therapy-System-level reliability testing is a complex task that involves evaluating and demonstrating the reliability of all the components of the system working together to perform the Therapy-System-Level mission task(s). A well-designed Therapy System Reliability Test should incorporate factors from design, human factors, manufacturing, service, and real-life therapy operations in the field. These factors include environmental conditions, human and end user interaction and programming, disposable and single use elements loading/removal, maintenance and cleaning, etc. A well-designed Therapy System Reliability Test should attempt to expose user interaction and system integration issues/failures not seen during standalone component or subsystem testing. Identified issues and failures provide valuable feedback to Human Factors, Manufacturing, and Service before the product is released to the field. Unlike what is often accepted or assumed in component reliability demonstration testing, Therapy system testing should not be designed or expected to be failure-free.

2. Therapy System Level Reliability Test Design

The Therapy System-level Reliability test design had multiple objectives listed as follows:

1. Demonstrate the targeted or required Reliability Metric can be achieved (e.g., Mean Time between Failure (MTBF), Mean-Cumulative Failures (MCF), etc.).
2. Discover system interactions, boundary issues and system sensitivities under laboratory conditions
3. Provide objective evidence that the system can maintain the reliability target over the expected service life (ESL) of the product.
4. Provide information on top failure modes and limited life components and inform or confirm the development of proactive maintenance or service strategies, if needed.
5. Identify Use/User errors and Human Factors related issues such as misuse, physical damage, etc.
6. Provide feedback to Service and Manufacturing functions based on observations/issues/failures.

3. System Description

The test item is a typical home-used medical device operated by the patient with prescription information fed into the system via either manual entry, or prescription stored on USB read device. The System is also intended for use by trained clinicians, trained patients, and non-clinician caregivers. The system is also designed to be transportable. The system user interacts with the system vi the user interface for programming, data entry, therapy information monitoring, and observation and clearing alarms and errors. System is configured to deliver medication through a disposable set that is stored at the patient's house. The operation of the device requires the patient to connect the set and bags, and either program the device to run the prescription or download a prescription from a USB thumb. The system configuration is shown in Figure 1.

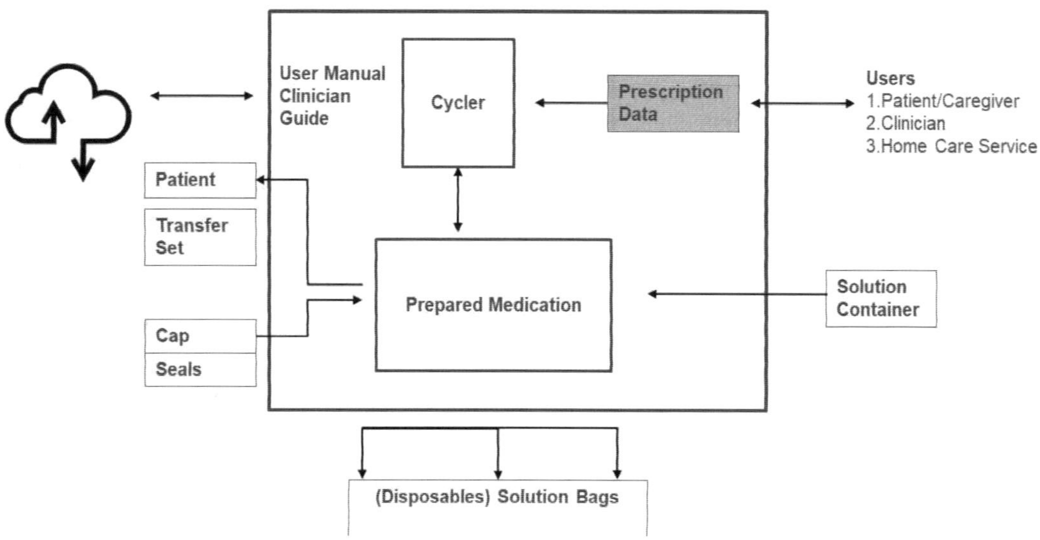

Figure 1. Medical Device System Configuration

4. Test Design Methodology Overview

The P-Diagram was constructed to define and list all types of signal, noise and control factors affecting the operation of the device. The design parameters were all allocated to control and noise factors in 5 main categories [1]:

- Piece to piece variation due to manufacturing defects and process control
- Change over time due to ageing of material or degradation due to usage and wear out
- Customer Usage in terms of misuse scenarios, handling, user error, cleaning, etc.
- External Environment such as during operation, shipping, and storage,
- System Interaction with disposable sets, patient loading and using and entering data into the device, disposable setting at patient's home, system competitivity to manufacturer database cloud, and service interface application.

Some of the critical design factors are used to design the test variables, activities, and measured outputs as illustrated in Figure 2. The test was conducted in an accelerated life context using multiple accelerants as follows:

1. Higher temperature
2. Manipulating system Software to change fluid flow rate & fill volume
3. 3x missions per day & reduced idle time (increased duty cycle)
4. Disposable bag height, therapy types
5. Other manual activation such as exercising AC/DC power, door actuation, loading and unloading of disposable elements, applying cleaning process, etc.

The repair and service was conducted per the standard service manual and following the standard troubleshooting procedures, calibration, etc.
Test results were analysed on a regular basis, and feedback provided to design, human factors, manufacturing, and service stakeholders for consideration of future design changes, service process enhancement, and manufacturing instructions updates and review.

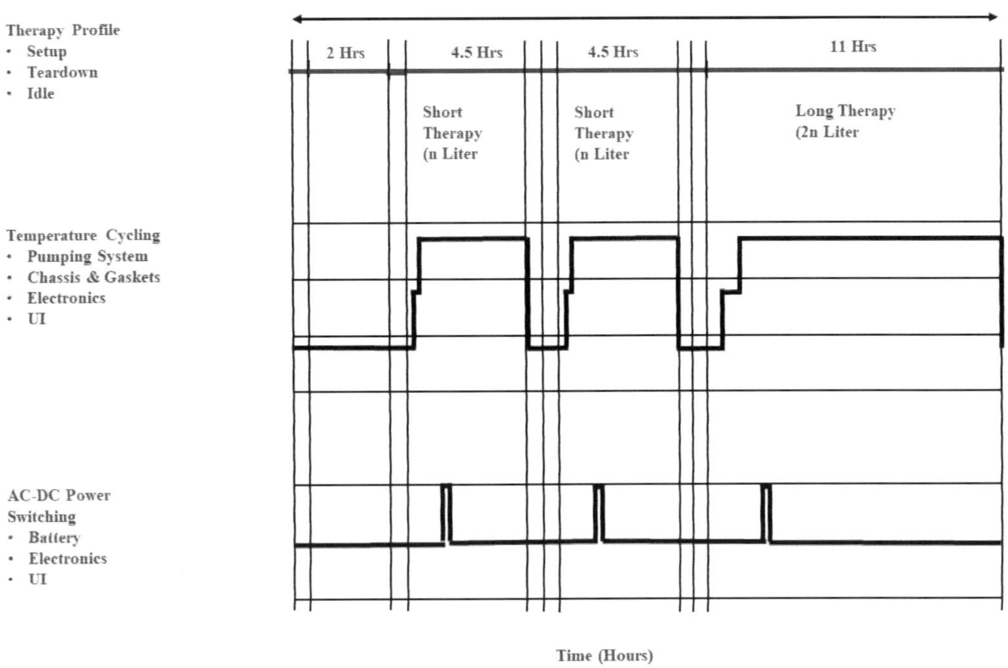

Figure 2. Combined Usage and Stress Cycling Profile

5. Preventive Maintenance (PM) Items

The Design and Service departments defined a list of limited life components and items that are prone to wear out and performance degradation over time. These items were listed for Proactive Service and maintenance (PM) based on time limits. In this test, PM components will not be replaced per the defined PM intervals, rather it will be tested until failure; and associated design parameters will be tracked and monitored for degradation over life. The objective of running these PM items to failure is to confirm and calculate the actual reliability of these items via actual time to failure of each. This is to understand if some expensive elements listed for PM remain within calibration limits; and may require just frequent recalibration instead of the scheduled replacement.

6. Test Sample Size

The sample size of the test was determined based on the reliability requirements, or target MTBF, which is interpreted into expected number of failure. Another factor considered was the fact that the MTBF will be monitored based on a N months rolling average value in the field after launch. System life testing should be designed to produce enough number of system failure to avoid overestimation of actual reliability. Use misuse, and wider usage profile spectrum allow for more issues and failures to be uncovered in the field, while never experienced during testing. We also know that electromechanical systems are complicated; and that they include many different modules and subsystems, along with a Software that is interacting with hardware and producing many errors due to triggers or thresholds violated. Published standards such as [2] indicates the testing time to achieve the target reliability metrics, such as MTBF. These standards indicate the testing time for specified number of f failures using Chi-Square distribution. They also indicate how the test time will change as the sample size is changing in the test, e.g., test time per sample decreases as the sample size increases. However, these standards don't recommend the proper sample size to produce enough number of failures to avoid the overestimation of the MTBF compared to real value realized in the field after launch. The sample size is determined to fulfill the minimum testing time with the maximum number of produced unique issues and errors, yet fulfilling the target MTBF.

7. System Acceleration Factor Calculation and Balance

The system acceleration factor achieved by the test setup and design will not be applied on all system modules. For each module, an individual acceleration Factor is calculated per the applicable failure mechanisms applicable to components in each module. Table 1 summarizes the applicable acceleration factors and rationale and algorithm used to calculate these individual acceleration factors:

Table 1. Modular Acceleration Factors for Different Accelerants & Application on Different Subsystems with Multiple Failure Mechanisms.

Subsystem / Component Category	Duty Cycle	Temp.	Power Cycle	Fill Volume - Flow Rate	Manual Actuation	No of Therapy - Missions	Cleaning
Electronic (PCB, Battery)	AF1	AF2	AF4			AF7	
Pumping System	AF1	AF2		AF5		AF7	
Moving Part (Door, etc.)	AF1				AF6	AF7	
Chassis / Enclosure:	AF1	AF3			AF6	AF7	AF8
User Interface (Keypad)	AF1	AF2			AF6	AF7	AF8
User Interface (Screen)	AF1	AF2	AF4			AF7	AF8
Heater-Heating System	AF1	AF2		AF5		AF7	

Where:

AF1. Ratio: test duty cycle / real life duty cycle for each subsystem

AF2. Most common model using acceleration factor calculated per temperature rise.

AF3. Aging of Polymer-based material; formulae and additional information are available in [3] and [4]

AF4, AF5, AF6, AF7, AF8: are linearly proportional to the rate of device usage, duty cycle & number of actuations, etc. Generic accelerants and other failure mechanisms can be found in other sources [5].

8. Conclusions & Recommendations

The following conclusions and recommendations are drawn from this study:

1. Defining the different control and noise factors is crucial in defining successful testing method that test for failure modes and simulate actual usage and operating environment.
2. Test sample size is crucial to accuracy and confidence in results to avoid error in actual MTBF.
3. Aging rate varies for different subsystems and components as acceleration factors in system life testing varies per the accelerant, failure mechanisms, and usage profile of each item.

9. Reference

[1] Juran and Gryna F 1993 Quality planning and analysis from product development through use *3rd ed.* (New York McGraw-Hill) p. 256

[2] MIL-HDBK-781A 1996 Reliability test methods, plans, and environments for engineering, development qualification, and production Washington DC

[3] ASTM D3045-92 2010 Standard Practice for Heat Aging of Plastics Without Load, ASTM International

[4] ASTM F1980-07 2011 Standard Guide for Accelerated Aging of Sterile Barrier Systems for Medical Devices

[5] Nelson W 2004 Accelerated Testing: Statistical Models, Test Plans, and Data Analysis (New York John Wiley & Sons) pp 521-548

IOP Conf. Series: Materials Science and Engineering **575** (2019) 012008 doi:10.1088/1757-899X/575/1/012008

Evaluation of Energy Harvesting Technique for Security Module

Y Nozaki[1*] and M Yoshikawa[1]

[1]Department of Information Engineering, Meijo University, 1-501 Shiogmaguchi, Tenpaku-ku, Nagoya, Aichi 468-8502, Japan

*143430019@ccalumni.meijo-u.ac.jp

Abstract. In internet of things (IoT), devices connect to external network in order to communicate data each other, and the improvement of human life is expected. However, IoT devices have problems to be solved, including design of low power and security such as authentication of devices. To overcome these problems, energy harvesting and physical unclonable function (PUF) have attracted attention. Actually, we have proposed a PUF using energy harvesting called energy harvesting PUF. However, since input of the energy harvesting PUF is fixed, available information for authentication is limited. Therefore, this study evaluates the energy harvesting PUF by increasing input value. Experiments using an actual energy harvester also showed the validity of the energy harvesting PUF.

1. Introduction

Energy harvesting technique, which harvests various energy and converts them to electric power, has attracted attention [1]–[4]. Energy harvesting is expected to apply for IoT devices because they have constraints including design of low power. By using energy harvesting technique, a system, which does not need battery replacement, can be developed [3]. Thus, energy harvesting can improve the energy efficiency of IoT devices.

IoT devices have problems to be solved [5]. Authentication of IoT devices is one of the problems. Generally, authentication utilizes an encryption technique using a secret key stored in a non-volatile memory on devices. However, it has been reported that the secret key is vulnerable against invasive attacks [6] and side-channel attacks [7], and countermeasures in high cost are required. Therefore, physical unclonable functions (PUFs) [8]–[10] have been proposed in order to ensure security of IoT devices in low cost. PUF is a circuit which extracts variations of semiconductor manufacturing as a device's ID. The secret key generated by using PUF does not need to store in a non-volatile memory. As a result, it has the resistance against illegal attacks. Then, a low cost authentication method using PUF, which does not utilize encryption processing, has also been proposed [9].

We have proposed a PUF using energy harvesting called an energy harvesting PUF [11]. In energy harvesting PUF, a small variance of power generation time due to production dispersion is used for device's authentication. The energy harvesting PUF can be also used in more low cost because it does not need to implement a dedicated PUF circuit additionally. On the other hand, in paper [11], input utilized in the energy harvesting PUF is fixed, and the generated unique value has only one type. Consequently, available information for authentication is limited, and a part of devices fails in authentication [11].

IOP Conf. Series: Materials Science and Engineering **575** (2019) 012008 doi:10.1088/1757-899X/575/1/012008

Therefore, this study evaluates the energy harvesting PUF by increasing input value. By experiments using an actual energy harvester, the validity of the energy harvesting PUF is evaluated.

2. Preliminaries

2.1. Energy Harvester

Figure 1 shows an energy harvester used in this study. As shown in figure 1, the energy harvester consists of a TWELITE-DIP, a power management module TWE-EH SOLAR, and a solar cell. In this energy harvester, when the power generation is finished, a TWELITE microcontroller is started. Then, wireless data communication is performed. Also, in this study, a MONOSTICK, including a TWELITE microcontroller, is used as a receiver.
In the TWEILTE microcontroller, the generated power by the photovoltaic generation is charged to a capacitor built in the TWE-EH SOLAR. At this time, when a voltage of the capacitor reaches 2.9 [V], the TWELITE microcontroller turns on. Then, wireless data communication is performed. When the voltage reaches below 2.0 [V], the TWELITE microcontroller turns off [12].

2.2. Related Works

First, E. Aponte has studied that the applying of energy harvesting for PUF [13]. In [13], the variance of the open-circuit voltage for each energy harvester, including solar cell and thermoelectric generator, was verified. Then, paper [13] showed the variance of the each energy harvester. However, in paper [13], methods for authentication of devices have not been studied.
Next, we have proposed energy harvesting PUF which uses the variance of the power generation time due to the production variation [11]. We have also shown that energy harvesting PUF can authenticate each device except for a part of devices. The detail of energy harvesting PUF is explained in section 3. Also, S. D. Kumar et al. have proposed a solar cell based PUF for secure key generation [14]. The solar cell based PUF consists of 8 solar cells and a Tiva TM4C123GH6PM microcontroller, and it generates a 128-bit ID from the measured open-circuit voltage by using an analog to digital converter (ADC). In addition, in [14], evaluations of randomness and reliability were also performed. However, authentication between devices has not been studied. Then, the solar cell based PUF needs as much as 8 solar cells; therefore, cost increases.

3. Energy Harvesting PUF

Energy harvesting PUF utilizes the production variation of the power generation time with energy harvester [11]. The generated power by the photovoltaic generation can be calculated by

$$P_{\max} = V_{OC} I_{SC} ff, \qquad (1)$$

where V_{OC} is the open-circuit voltage, I_{SC} is the short-circuit current, and *ff* is the fill factor.
In addition, the open circuit voltage V_{OC} can be represented by

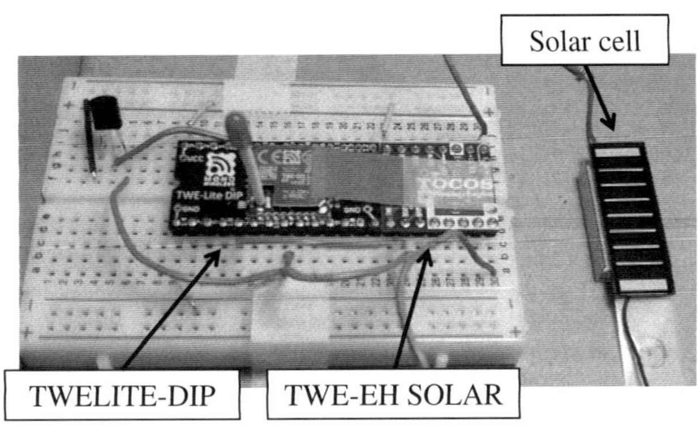

Figure 1. Energy harvester.

$$V_{OC} = \frac{kT}{q}\ln\left(\frac{(N_A + \Delta n)\Delta n}{n_i^2}\right), \tag{2}$$

where kT/q is the thermal voltage, N_A is the doping concentration, Δn is the excess carrier concentration, and n_i^2 is the intrinsic carrier concentration [15][16].

Since the doping concentration of each semiconductor is different due to the production variation, the open-circuit voltage V_{OC}, which is depend on N_A (see formula (2)), is different [13]; therefore, the generated power differs. As a result, the power generation time is changed in each energy harvester [11]. The energy harvesting PUF utilizes the power generation time for device's authentication. Figure 2 shows the outline of the energy harvesting PUF. Here, the energy harvester used in this study communicates data when the voltage of the capacitor is charged to reach 2.9 [V] (see subsection 2.1). Then, when the voltage reaches below 2.0 [V], the TWELITE microcontroller turns off, and the charging of the power generation is carried out again. The energy harvesting PUF calculates the power generation time between the data receiving times t_1 and t_2. At this time, t_1 is the receiving time when the TWEILTE turns on, and t_2 is the receiving time before the TWELITE turns off.

Specifically, the power generation time T can be calculated by

$$T = t_2 - t_1. \tag{3}$$

For the authentication, first, the power generation time $T_{register}$ is registered to the database. Then, the Euclid distance d between the power generation time during the authentication (T_{user}) and that during the registration ($T_{register}$). This can be calculated by

$$d = \left|T_{register} - T_{user}\right| = \sqrt{(T_{register} - T_{user})^2}. \tag{4}$$

Then, the Euclid distance, whose value is the minimum, is authenticated as a correct energy harvester. On the other hand, in [11], since light incident position as input is fixed, available information for authentication is limited. Therefore, it has been pointed out that a part of devices fails in authentication [11].

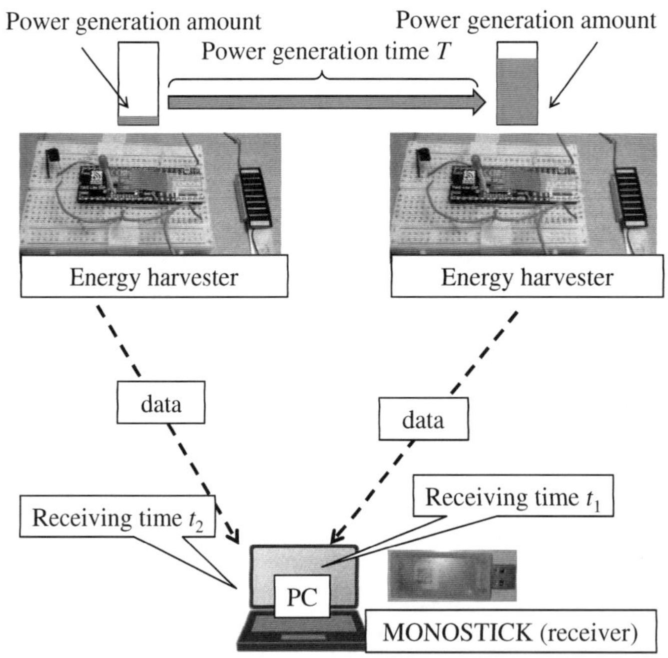

Figure 2. Outline of the energy harvesting PUF.

4. Experiments

This study evaluates the energy harvesting PUF by increasing input value to improve the authentication accuracy.

4.1. Experimental Environment

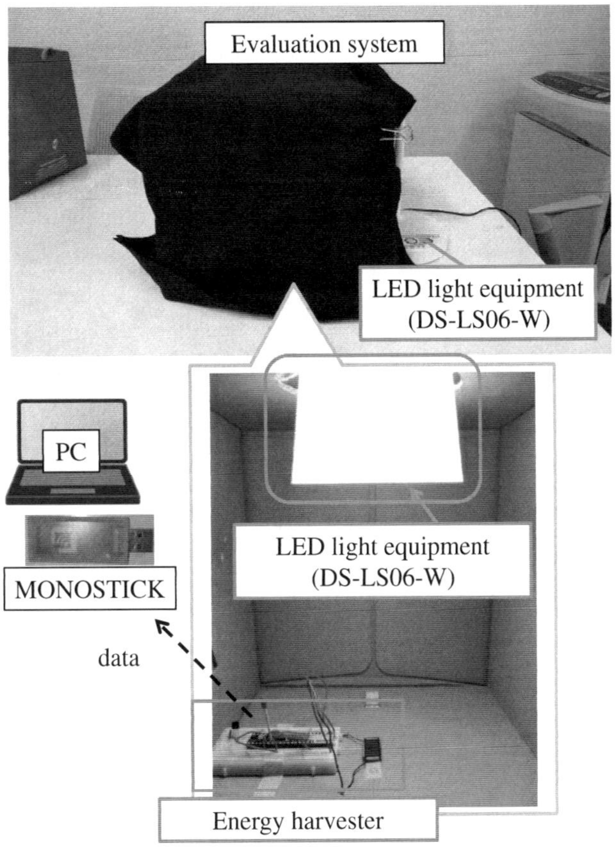

Figure 3. Evaluation system.

Table 1. Experimental condition.

Energy harvester	TWELITE-DIP
	TWE-EH SOLAR
	Solar cell (Panasonic AM-5815)
LED light equipment	DL-LS06-W
Receiver	MONOSTICK

Figure 3 and table 1 show the experimental environment. Experiments used an evaluation system covered with black cloth to reduce the influence of light from the outside. An energy harvester, which consists of a TWELITE microcontroller (TWELITE-DIP), a power management module (TWE-EH SOLAR), and a solar cell, was placed in the evaluation system. In addition, LED light equipment (DS-LS06-W) was set on the energy harvester, and it irradiated constant light to the solar cell. At this time, the illuminance of the LED light equipment is 1,000 [lx] against directly under 25cm. Then, to

IOP Conf. Series: Materials Science and Engineering **575** (2019) 012008 doi:10.1088/1757-899X/575/1/012008

measure the power generation time, constant light was irradiated into 3 types of solar cells (A, B, and C). For the measurement of the power generation time, a MONOSTICK was used as a receiver.
To utilize additional information for authentication, experiments by changing of light incident position was conducted. Here, by changing the device's position, the light incident position can be changed. Specifically, experiments utilized 3 types of device's positions: middle, bottom, and top. Figure 4 shows the device's positions.

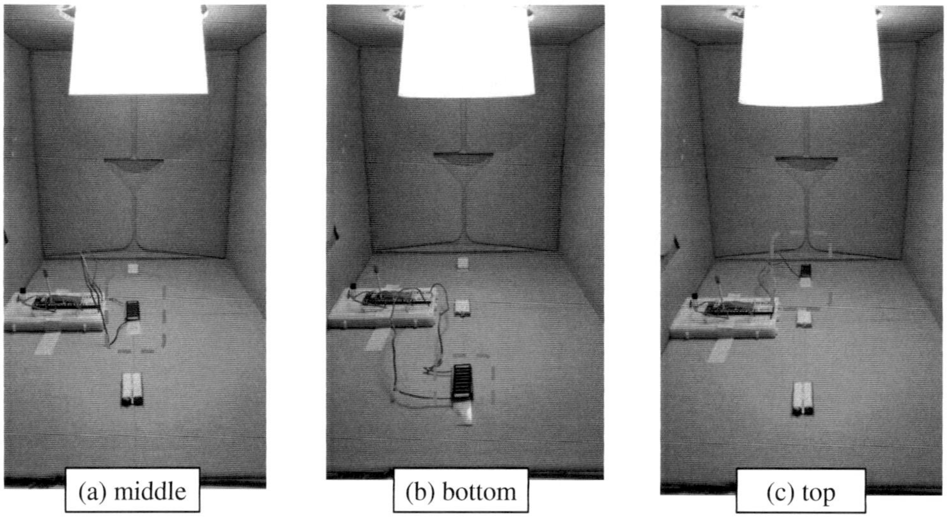

Figure 4. Device's positions.

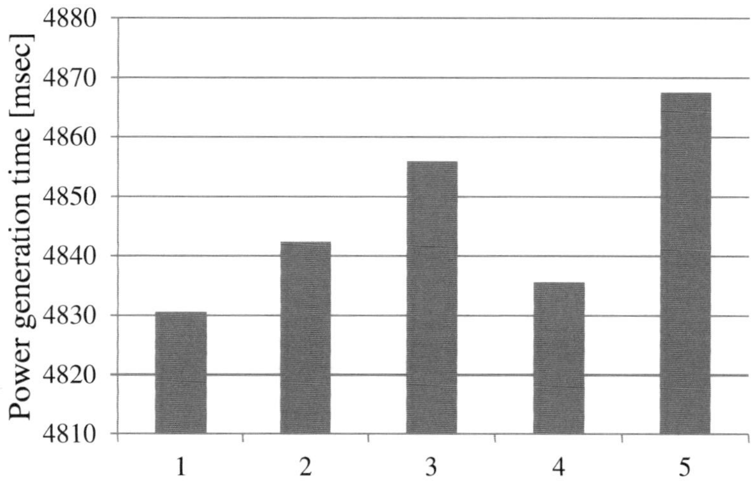

Figure 5. Results of preliminary experiments.

4.2. Experimental Results
First, the steadiness of energy harvesting PUF was verified as preliminary experiments. Experiments measured the average power generation time of a same solar cell for 5 times. Figure 5 shows the verification results. In this figure, the vertical line shows the average power generation time. Then, the difference between the minimum (the first of figure 5) and the maximum (the fifth of figure 5) of the power generation time was calculated by the Euclid distance. It was confirmed that the calculated value was 37.

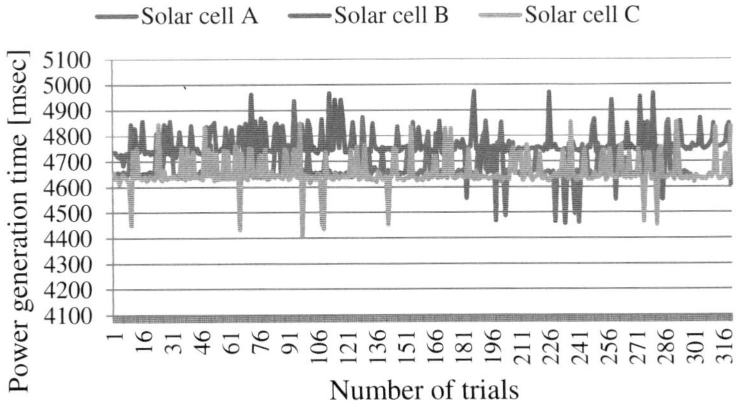

Figure 6. Results of the middle position.

Table 2. Comparison results (middle).

	A	B	C
A	—	93.92	15.49
B	93.92	—	109.4
C	15.49	109.4	—

Table 3. Comparison results (bottom).

	A	B	C
A	—	350.8	164.1
B	350.8	—	186.6
C	164.1	186.4	—

Table 4. Comparison results (top).

	A	B	C
A	—	176.4	44.77
B	176.4	—	131.7
C	44.77	131.7	—

Next, experiments verified the variance of the power generation time similar to paper [11], which placed solar cells in one type of position (middle position). Figure 6 shows the experimental results. In this figure, the vertical line shows the power generation time. In addition, table 2 shows differences of average power generation time between solar cells. The values of table 2 were calculated by the Euclid distance. As shown in figure 6 and Table 2, the power generation time of solar cell B differs from other solar cells (A and C) due to the production variation. By contrast, the power generation time of solar cell A and C is very similar, and its difference is 15.49 (see table 2). As discussed in the previous paragraph, the difference of the power generation time in the same solar cell is 37 in the maximum. Therefore, authentication between solar cells A and C is difficult because its difference is less than 37. Thus, authentication of a part of devices may fail when device's position is fixed to only one.

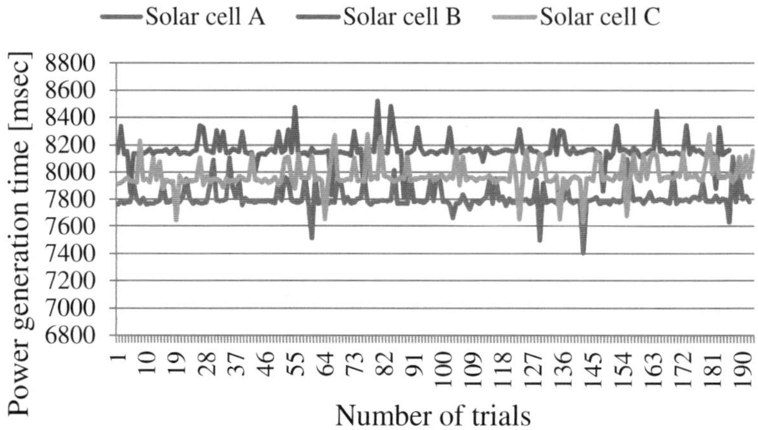

Figure 7. Results of the bottom position.

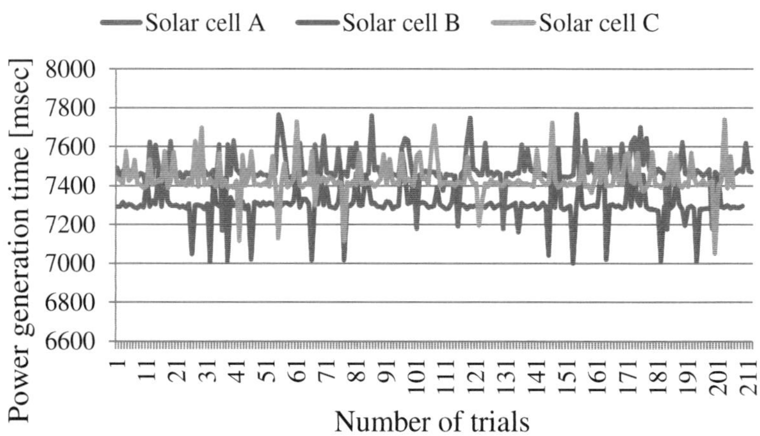

Figure 8. Results of the top position.

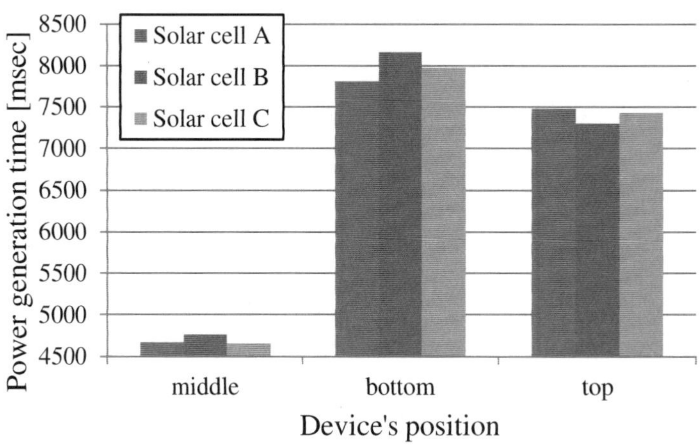

Figure 9. Comparison of average power generation time in each position.

Finally, experiments by changing the device's position were conducted. Figures 7 and 8 show the experimental results. Figure 7 is the results of the bottom position and figure 8 is that of the top position. Moreover, tables 3 and 4 show the results of average power generation time, and figure 9 shows the comparison results. As shown in these figures, in particular, the power generation time between solar cells A and C differs compared with that of the middle position. For example, on the bottom position, the power generation time of solar cell A is larger than that of solar cell C (difference is 164). This is presumably because the power generation efficiency is different in each device by different light incident position. Therefore, by using more information not only in the middle position but also in the bottom and the top position, the authentication accuracy can be improved.

5. Conclusion

This study evaluated the energy harvesting PUF by increasing input value. In this study, to utilize more input values for the energy harvesting PUF, the light incident position was changed. For changing of the light incident position, 3 types of device's positions (middle, bottom, and top) were used. Experiments using 3 types of solar cells (A, B, and C) showed that when the device was fixed in middle position, the difference of power generation time between solar cells A and C was less than 37 which is the variance of a same solar cell. Thus, this study clarified that authentication fails when input value is fixed. In addition, experiments also showed that when the device was fixed in bottom, the power generation time of solar cell A was larger than that of solar cell C (its difference was 164), that is, the authentication succeeded. Thus, the authentication accuracy could be improved by using additional information not only in the middle position but also in the bottom and top position.

Future works include that the direction of solar cells is changed in order to use more information for energy harvesting PUF, and the study of other energy harvesters.

6. References

[1] Chalasani S and Conrad J M 2008 A Survey of Energy Harvesting Sources for Embedded Systems *Proc. IEEE SoutheastCon* pp 442–447

[2] Nicosia A, Pau D, Giacalone D, Plebani E, Bosco A and Iacchetti A 2018 Efficient Light Harvesting for Accurate Neural Classification of Human Activities *Proc. IEEE Int. Conf. on Consumer Electronics* pp 1–4

[3] Tentzeris M M, Georgiadis A and Roselli L 2014 Energy Harvesting and Scavenging *Proc. IEEE* **102** 11 pp 1644–1648

[4] Dell'Anna F, Dong T, Li P, Yumei W, Yang Z, Casu M R, Azadmehr M and Berg Y 2018 State-of-the-Art Power Management Circuits for Piezoelectric Energy Harvesters *IEEE Circuits and Systems Magazine* **18** 3 pp 27–48

[5] Sicaria S, Rizzardia A, Griecob L A and Coen-Porisinia A 2015 Security, privacy and trust in Internet of Things: The road ahead *Computer Networks* **76** pp 146–164

[6] Skorobogatov S P 2015 Semi-Invasive Attacks – A New Approach to Hardware Security Analysis *PhD thesis University of Cambridge*

[7] Kocher P, Jaffe J and Jun B 1999 Differential Power Analysis *Proc. CRYPTO'99* LNCS 1666 pp 388–397

[8] Gassend B, Clarke D E, Dijk M V and Devadas S 2002 Silicon Physical Random Functions *Proc. 18th Annual Computer Security Applications Conf.* pp 148–160

[9] Aman M N, Chua K C and Sikdar B 2016 Physical Unclonable Functions for IoT Security *Proc. 2nd ACM Int. Workshop on IoT Privacy, Trust, and Security* pp 10–13

[10] Marchand C, Bossuet L, Mureddu U, Bochard N, Cherkaoui A and Fischer V 2018 Implementation and Characterization of a Physical Unclonable Function for IoT: A Case Study With the TERO-PUF *IEEE Trans. Computer-Aided Design of Integrated Circuits and Systems* **37** 1 pp 97–109

[11] Nozaki Y and Yoshikawa M 2018 Physical Unclonable Function using Energy Harvesting *Proc. 2018 IEEE Int. Meeting for Future of Electron Devices, Kansai* pp 38–39

[12] https://mono-wireless.com/jp/index.html

[13] Aponte E 2017 A Study on Energy Harvesters for Physical Unclonable Functions and Random Number Generation *M.S. thesis Virginia Polytechnic Institute and State University.*
[14] Kumar S D, Labrado C, Badhan R, Thapliyal H and Singh V 2018 Solar Cell Based Physically Unclonable Function for Cybersecurity in IoT Devices *Proc. 2018 IEEE Computer Society Annual Symp. VLSI* pp 697–702
[15] Sinton R A, Cuevas A and Stuckings M 1996 Quasi-Steady-State Photoconductance, A New Method for Solar Cell Material and Device Characterization *Proc. 25th IEEE Photovoltaic Specialists Conference* pp 457–460
[16] Sinton R A and Cuevas A 2000 A Quasi-Steady-State Open-Circuit Voltage Method for Solar Cell Characterization *Proc. 16th European Photovoltaic Solar Energy Conf.* pp 1152–1155

Acknowledgments
This work was supported by JSPS KAKENHI Grant Number 17J11408.

IOP Conf. Series: Materials Science and Engineering **575** (2019) 012009 doi:10.1088/1757-899X/575/1/012009

Extended Order High Gain Observer Based Stabilization of 2 DOF Pan Tilt Platform for Aerial Imaging System

O Altaf[1], F M Malik[1], R A Azim[2], S Hussain[1] and N Ahmad[1]

[1]National University of Sciences and Technology, DEE CE&ME, Islamabad, Pakistan
[2]National University of Sciences and Technology, DME CE&ME, Islamabad, Pakistan

omair.altaf1@gmail.com

Abstract. Aerial imaging systems are mounted on two axis serial robotic manipulators with high powered pan and tilt actuations, known as a Gimbal. The angular displacement of the gimbal is measured through very high precision sensors, typically shaft encoders with high PPR specifications. The purpose of the gimbal is to keep the orientation of the payload towards the target irrespective of the displacement of the UAV (unmanned aerial vehicle) and any other vibrations/ disturbances that might be introduced into the system. Because of 2-DOF (degrees of freedom) dynamical model of the gimbal, frictional force, noise, cable restraint, disturbances from the external environment along with the motions of the vehicle's body (as a result of manoeuvring or vibration), the spotting and tracking accuracy of the gimbal platform system may significantly degrade. The nature of these redundant disturbances is mostly of nonlinear nature and their modelling is a difficult task as they keep on changing. In order to cancel out these redundant disturbances and uncertainties, extended order high gain observer (EHGO) based feedback linearization control is used and extensive simulations are performed to show the comparison of tracking control without EHGO and after augmenting EHGO.

1. Introduction

Inertial stabilization platforms (ISPs) have various engineering applications in modern world like these platforms can be used as housing for military weapons, astronomical equipment and cameras. Designs of these systems are configured according to the requirements, applications and desired performance [1]. The primary objective of these models is to mitigate the effects of aerial manoeuvres, platform motions, engine vibrations and nonlinear disturbances. All of these disturbances results in decreasing the accurate pointing capability of the ISPs [2].

The stability of imaging systems mounted on gimbal, on board UAVs is fundamental for accurate surveillance/ target, more so, for long range sensors/ target designators termed as payload. In this regard a robust nonlinear control scheme based on detailed mathematical model for the two axis gimbal of the UAV that houses the payload is the solution since the gimbal is subjected to numerous un-modelled disturbance which include and are not limited to parameter uncertainties, wind gusts, engine vibrations etc.

The paper includes mathematical modelling of the 2-axis gimbal, nonlinear tracking controller design for it and simulations of the system with its capability to reject unwanted disturbances.

Paper is divided into following subsections, mathematical modelling is defined in section 2 followed by feedback linearization in section three. Section 4 explains extended order high gain observer design and simulation results are shown in section 5. Conclusion is given in section 6.

2. Mathematical Modelling

As is customary in the field of aerospace and robotics, quantification of moving elements is accomplished by attaching frames to links. In case of gimbal the base frame which is attached with the UAV is marked as frame 0. The motion of first rotary joint, termed as pan, is represented by frame 1. The first joint is also named as outer gimbal in some texts. The second joint has frame 2 attached with it. It is pertinent to note here that the first link rotates about the z_1 axis by an angle θ_1 (pan angle) and the second link rotates about the y_2 axis by and angle θ_2 (tilt angle). The affixing of frames provides the foundation for dynamic modelling of the gimbal. [3, 4].

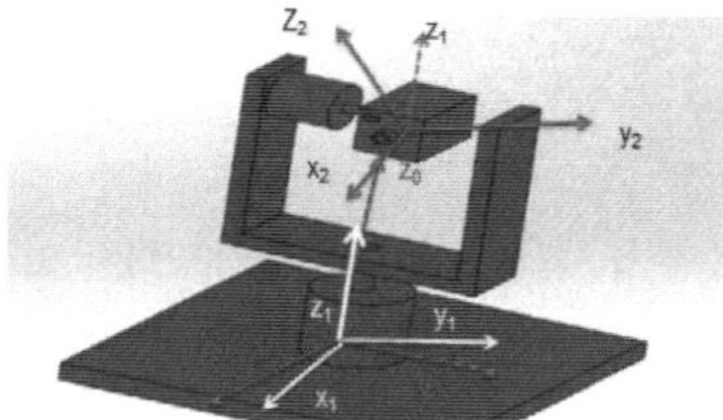

Figure 1. Schematic model of a 2 DOF Gimbal

Mathematical modelling of a system is a preliminary step for control design. As the 2-axes gimbal is a two link serial robotic manipulator, its modelling is done in the classical style of robotic manipulators. The method is well-known "Iterative Newton Euler Dynamics". [3].

The dynamic model of 2-axes gimbal is evolved using the kinematic and dynamic expressions listed above. The kinematic expressions are evolved from base frame to link 2. The base frame is considered static for a slowly maneuvering aircraft [3].

2.1 Base Frame

Linear velocity, angular velocity, linear acceleration, angular acceleration turns out to be

$$\begin{aligned} {}^{0}\upsilon_0 &= [0,0,0]^T \;, {}^{0}\omega_0 = [0,0,0]^T \\ {}^{0}\dot{\upsilon}_0 &= [0,0,0]^T , {}^{0}\dot{\omega}_0 = [0,0,0]^T \end{aligned} \tag{1}$$

2.2 Inner Gimbal

The rotation matrix of frame 1 with respect to base frame is given by

$${}^{0}_{1}R = \begin{bmatrix} c_1 & -s_1 & 0 \\ s_1 & c_1 & 0 \\ 0 & 0 & 1 \end{bmatrix} \tag{2}$$

Linear velocity, angular velocity, linear acceleration and angular acceleration of frame 1 are given below

$$\begin{aligned} {}^{1}\upsilon_1 &= [0\ 0\ 0]^T , {}^{1}\dot{\upsilon}_1 = [0\ 0\ 0]^T \\ {}^{1}\omega_1 &= [0\ 0\ \dot{\theta}_1]^T , {}^{1}\dot{\omega}_1 = [0\ 0\ \ddot{\theta}_1]^T \end{aligned} \tag{3}$$

2.3 Outer Gimbal

The outer gimbal is serially connected with the inner gimbal. The outer gimbal rotates about its y axis. The rotation matrix of the outer gimbal with respect to the inner gimbal is given as:

$${}^{1}_{2}R = \begin{bmatrix} c_2 & 0 & s_2 \\ 0 & 1 & 0 \\ -s_2 & 0 & c_2 \end{bmatrix} \tag{4}$$

The linear velocity, angular velocity, linear acceleration and the angular acceleration of the outer gimbal are expressed as:

$$\begin{aligned} &{}^{2}\upsilon_2 = [0\ 0\ 0]^T,\ {}^{2}\dot{\upsilon}_2 = [0\ 0\ 0]^T \\ &{}^{2}\omega_2 = [-s_2\dot{\theta}_1\ \ \dot{\theta}_2\ \ c_2\dot{\theta}_2]^T \\ &{}^{2}\dot{\omega}_2 = [-s_2\ddot{\theta}_1 - c_2\dot{\theta}_1\dot{\theta}_2\ \ \ddot{\theta}_2\ \ c_2\ddot{\theta}_1 - s_2\dot{\theta}_1\dot{\theta}_2]^T \end{aligned} \tag{5}$$

According to the Newton's laws the resultant force acts on the centre of gravity (CG) of a rigid body. This warrants locating the CG of each link. The CG of the inner and outer gimbals expressed with respect to their frames of rotation is given as follows.

$${}^{1}p_{C_1} = \begin{bmatrix} 0 \\ 0 \\ z_{c1} \end{bmatrix},\ {}^{2}p_{C_2} = \begin{bmatrix} 0 \\ 0 \\ z_{c2} \end{bmatrix}$$

Similarly the Euler equations for rigid bodies require inertia matrices. The inertia matrices for the two gimbals are presented as follows:

$${}^{c1}I_1 = \begin{bmatrix} I_{1x} & 0 & 0 \\ 0 & I_{1y} & 0 \\ 0 & 0 & I_{1z} \end{bmatrix},\ {}^{c2}I_2 = \begin{bmatrix} I_{2x} & 0 & 0 \\ 0 & I_{2y} & 0 \\ 0 & 0 & I_{2z} \end{bmatrix}$$

The evaluation of kinematic equations for the 2-axes gimbal leads to the expressions for resultant forces and moments on each link. The resultant forces on inner and outer gimbals are expressed as:

$${}^{1}F_1 = {}^{2}F_2 = [0\ 0\ 0]^T \tag{6}$$

The expressions for resultant moment on gimbal is given by

$${}^{1}N_1 = [0\ 0\ J_{1z}\ddot{\theta}_1]^T \tag{7}$$

Similarly the expression for net moment on the outer gimbal is as follows

$${}^{2}N_2 = \begin{bmatrix} \left(-s_2\ddot{\theta}_1 - c_2\dot{\theta}_1\dot{\theta}_2\right)J_{2x} + c_2\dot{\theta}_1\dot{\theta}_2 J_{2z} - c_2\dot{\theta}_1\dot{\theta}_2 J_{2y} \\ \ddot{\theta}_2 J_{2y} - c_2 s_2 \dot{\theta}_1^2 J_{2x} + c_2 s_2 \dot{\theta}_1^2 J_{2z} \\ \left(c_2\ddot{\theta}_1 - s_2\dot{\theta}_1\dot{\theta}_2\right)J_{2z} + s_2\dot{\theta}_1\dot{\theta}_2 J_{2x} - s_2\dot{\theta}_1\dot{\theta}_2 J_{2y} \end{bmatrix} \tag{8}$$

2.4 Dynamical modeling

The next step in dynamic modeling [3] of the gimbal is inward iteration for calculation of moments applied by link 1 (inner gimbal) on link 2 (outer gimbal) and by base on link. The moment applied by inner gimbal on outer gimbal is given as:

$${}^{2}n_2 = \begin{bmatrix} -\left(\ddot{\theta}_1 s_2 + \dot{\theta}_1\dot{\theta}_2 c_2\right)I_{2x} - \dot{\theta}_1\dot{\theta}_2 c_2 I_{2y} + \dot{\theta}_1\dot{\theta}_2 c_2 I_{2z} \\ -\dot{\theta}_1^2 c_2 s_2 I_{2x} + \ddot{\theta}_2 I_{2y} + \dot{\theta}_1^2 c_2 s_2 I_{2z} \\ \dot{\theta}_1\dot{\theta}_2 s_2 I_{2x} - \dot{\theta}_1\dot{\theta}_2 s_2 I_{2y} + \left(\ddot{\theta}_1 c_2 - \dot{\theta}_1\dot{\theta}_2 s_2\right)I_{2z} \end{bmatrix} \tag{9}$$

The second row represents the rotary joint about which rotation can take place. This rotation is caused by the actuator torque τ_2. Equating the second row with the actuator torque and rearranging leads to the following expression which is the desired dynamic equation for the second joint.

$$\ddot{\theta}_2 = \frac{\tau_2 + \dot{\theta}_1^2 c_2 s_2 I_{2x} - \dot{\theta}_1^2 c_2 s_2 I_{2z}}{I_{2y}} \tag{10}$$

Proceeding inwards further, the moment applied by the base frame on the inner gimbal is calculated and after evaluating, the following dynamic equation for the first joint is obtained [3].

$$\ddot{\theta}_1 = \frac{\tau_1 - 2\dot{\theta}_1\dot{\theta}_2 c_2 s_2 I_{2x} + 2\dot{\theta}_1\dot{\theta}_2 c_2 s_2 I_{2z}}{I_{1z} + s_2^2 I_{2x} + c_2^2 I_{2z}} \tag{11}$$

3. Feedback Linearization

Feedback linearization is a control technique used for controlling nonlinear systems by choosing an appropriate control input and transforming the nonlinear system into a linearized system [6, 7]. In order to linearize the system model, first it should be written in state space form.

3.1 State space model

The two joint angles and their respective velocities are chosen as state variable as follows:

$$x_1 = \theta_1, x_3 = \theta_2, u_1 = \tau_1, u_2 = \tau_2$$

The state space representation of the 2-axis gimbal model is given by

$$\begin{aligned}
\dot{x}_1 &= x_2 \\
\dot{x}_2 &= \frac{(-2J_{2x}\cos x_3 \sin x_3 . x_2 x_4 - 2J_{2z}\cos x_3 \sin x_3 . x_2 x_4) + u_1}{J_{1z} + J_{2z}(\cos x_3)^2 + J_{2x}(\sin x_3)^2} \\
\dot{x}_3 &= x_4 \\
\dot{x}_4 &= \frac{(J_{2z}\sin x_3 \cos x_3 .(x_2)^2 - J_{2x}\sin x_3 \cos x_3 .(x_2)^2) + u_2}{J_{2y}}
\end{aligned} \tag{12}$$

3.2 Control Design

The next step towards designing EHGO is control design [8, 9].

The control objective for the 2-axes gimbal is the tracking of desired pan and tilt angles for the gimbal. The desired angles are computed by the geo-pointing algorithm on the basis of target location, UAV location and UAV attitude provided by the on board gyros. The desired reference angles for pan and tilt are respectively termed as r_1, r_2. The control objective is to asymptotically track these angles which are mathematically defined as

$$\lim_{t\to\infty}(\theta_1 - r_1) \to 0, \quad \lim_{t\to\infty}(\theta_2 - r_2) \to 0 \tag{13}$$

In order to achieve tracking the state space representation transformed into error coordinates is

$$\begin{aligned}
e_1 &= x_1 - r_1 \\
e_2 &= x_2 - \dot{r}_1 \\
e_3 &= x_3 - r_2 \\
e_4 &= x_4 - \dot{r}_2
\end{aligned} \tag{14}$$

The error dynamics are given as

$$\begin{aligned}
\dot{e}_1 &= e_2 \\
\dot{e}_2 &= f_1(x) + g_1(x)u_1 - \ddot{r}_1 \\
\dot{e}_3 &= e_3 \\
\dot{e}_4 &= f_2(x) + g_2(x)u_2 - \ddot{r}_2
\end{aligned} \tag{15}$$

where for convenience of notation, the nonlinear terms are defined as

$$\begin{aligned}
f_1(x) &= \frac{2(\sin x_3 \cos x_3 . x_2 . x_4).(J_{2z} - J_{2x})}{J_{1z} + J_{2z}\cos x_3^2 + J_{2x}\sin x_3^2} \\
f_2(x) &= \frac{(\sin x_3 \cos x_3 . x_2^2).(J_{2z} - J_{2x})}{J_{2y}} \\
g_1(x) &= \frac{1}{J_{1z} + J_{2z}\cos x_3^2 + J_{2x}\sin x_3^2} \\
g_2(x) &= \frac{1}{J_{2y}}
\end{aligned} \tag{16}$$

Tracking is achieved by the following state feedback control law which is based on feedback linearization technique [5,6]

$$\begin{aligned}
u_1 &= g_1^{-1}(x)\left(-f_1(x) + \ddot{r}_1 + v_1\right) \\
u_2 &= g_2^{-1}(x)\left(-f_2(x) + \ddot{r}_2 + v_2\right)
\end{aligned} \tag{17}$$

The above control input linearizes the error dynamics of the system, which are given by

$$\begin{aligned}
\dot{e}_1 &= e_2 \\
\dot{e}_2 &= v_1 \\
\dot{e}_3 &= e_4 \\
\dot{e}_4 &= v_2
\end{aligned} \tag{18}$$

Equation (3.7) is of a controllable linear system, the origin of which if stabilized, guarantees asymptotic tracking. To this effect following input is applied [7]

$$\begin{aligned}
v_1 &= 33.35e_1 + 11.58e_2 - 2.92e_3 - 0.51e_4 \\
v_2 &= -2.47e_1 - 0.47e_2 + 38.39e_3 + 11.41e_4
\end{aligned} \tag{19}$$

where the gains are selected such that the eigenvalues of the closed loop system are at -5, -5.5, -6, -6.5.

4. Extended Order High Gain Observer

In any practical control system, the availability of all system states is not possible, to account for the unmeasured system states, observers are used, for nonlinear systems like the 2-axes gimbal, a variant of the high gain observer, known as extended order high gain observer (EHGO) has been employed. The strength of the high gain observer is that it can estimate modelling uncertainties and disturbances in addition to system states [8]. The EHGO for the 2-axes gimbal is presented as follows [8, 9]

$$\begin{aligned}
\dot{e}_1 &= e_2 + (\frac{\alpha_1}{\varepsilon}).(e_1 - e_1) \\
\dot{e}_2 &= g_1.u_1 - \ddot{r}_1 + (\frac{\alpha_2}{\varepsilon^2}).(e_1 - e_1) + \sigma_1 \\
\dot{\sigma}_1 &= (\frac{\alpha_3}{\varepsilon^3}).(e_1 - e_1) \\
\dot{e}_3 &= e_4 + (\frac{\alpha_1}{\varepsilon}).(e_3 - e_3) \\
\dot{e}_4 &= g_2.u_2 - \ddot{r}_2 + (\frac{\alpha_2}{\varepsilon^2}).(e_3 - e_3) + \sigma_2 \\
\dot{\sigma}_2 &= (\frac{\alpha_3}{\varepsilon^3}).(e_3 - e_3)
\end{aligned} \tag{20}$$

The observer design parameters are given as ϵ=0.1, α_1=6, α_2=11, α_3=6. With the estimates of all perturbations made available by the EHGO, the control law is modified as:

$$\begin{aligned} u_1 &= g_1^{-1}(x)(-f_1(x)+\ddot{r}_1+v_1-\sigma_1) \\ u_2 &= g_2^{-1}(x)(-f_2(x)+\ddot{r}_2+v_2-\sigma_2) \end{aligned} \tag{21}$$

5. Simulation Results

The tracking control law is simulated for tracking constant and sinusoidal signals. Two different scenarios for perturbation are considered. In one case EHGO is not employed and it becomes evident that tracking error persists, in the other case when estimates of perturbations obtained from the EHGO are incorporated into the control law, perfect tracking is achieved. The tracking results for various scenarios are presented as follows

5.1 Tracking without EHGO

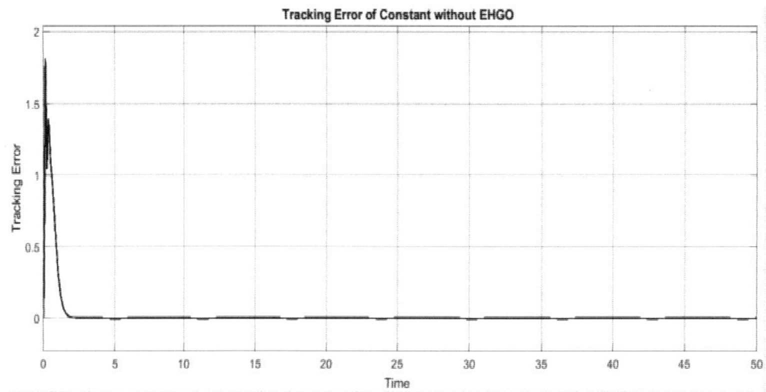

Figure 2. Tracking error of a constant without EHGO

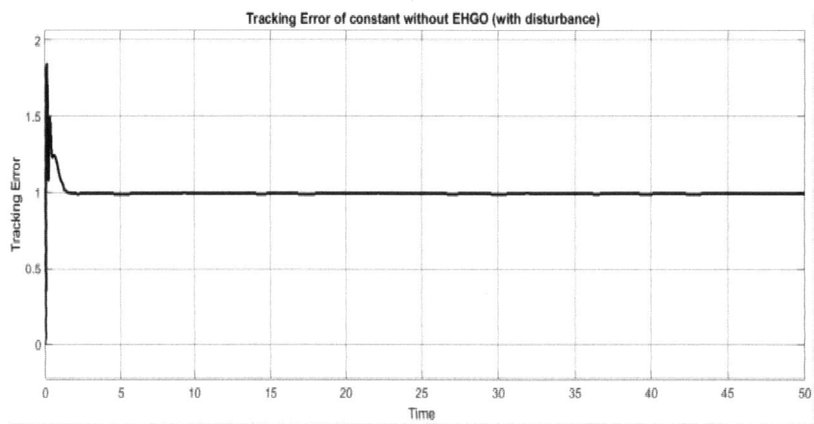

Figure 3. Tracking error of a constant without EHGO with disturbance

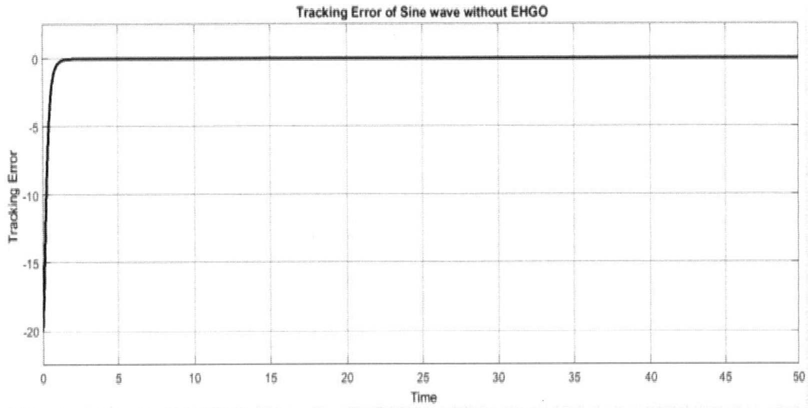

Figure 4. Tracking error of a sine wave without EHGO

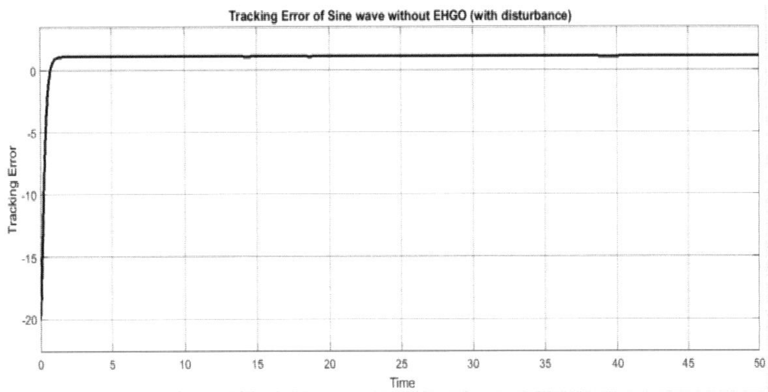

Figure 5. Tracking error of a sine wave without EHGO with disturbance

5.2 EHGO based tracking

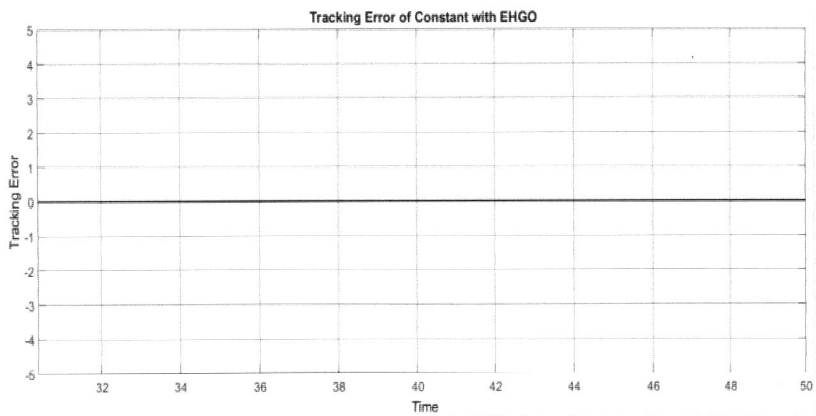

Figure 6. Tracking error of constant with EHGO

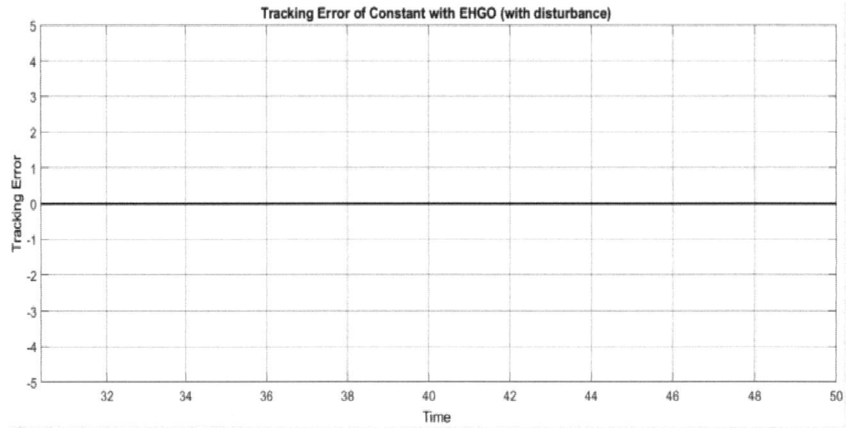

Figure 7. Tracking error of constant with EHGO (with disturbance)

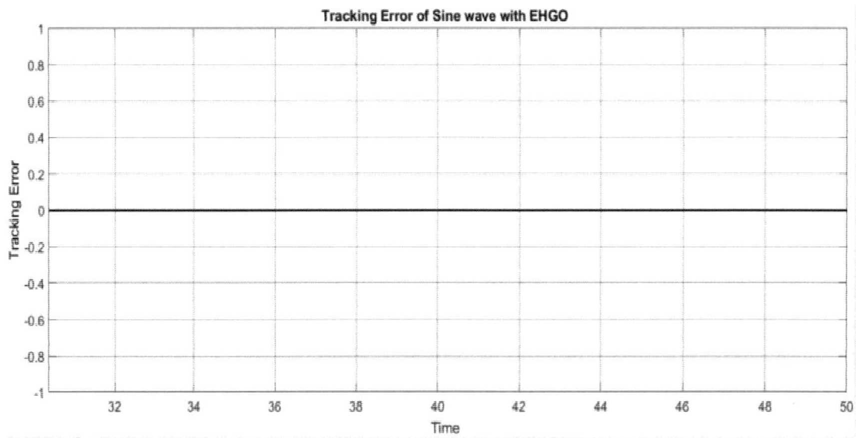

Figure 8. Tracking error of a sine wave with EHGO

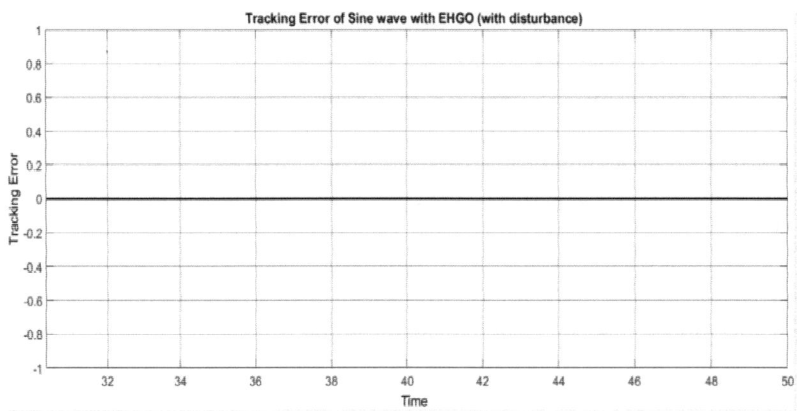

Figure 9. Tracking error of a sine wave with EHGO (with disturbance)

6. Conclusion

The simulations indicate the efficacy of the EHGO in estimating perturbations which are then used to achieve asymptotic tracking with feedback linearization tracking control law. The tracking of reference pan and tilt angles for the 2-axes gimbal is achieved.

IOP Conf. Series: Materials Science and Engineering **575** (2019) 012009 doi:10.1088/1757-899X/575/1/012009

7. References

[1] J. Hilkert. Inertially stabilized platform technology concepts and principles, IEEE Control Systems, ***28*(1)**, 26-46 ,(2008).

[2] P. J. Kennedy, R. L. Kennedy. *Direct Versus Indirect Line of Sight (LoS) Stabilization*. IEEE Transactions on Control Systems Technology, **11(1):**3–15, (2003).

[3] J.J Craig. *Introduction to Robotics: Mechanics & Control*, (1986)

[4] B. Ekstrand. Equations of Motion for A Two Axes Gimbal System, IEEE Transactions on Aerospace and Electronic Systems, **37(3)**:1083–1091, (2001).

[5] M.K. Masten, "Inertially Stabilized Platform for Optical Imaging Systems", IEEE Control Systems Magazine, February, (2008).

[6] A. Isidori, *Nonlinear Control Systems*, Springer, (1985)

[7] H. K Khalil, J. Grizzle, *Nonlinear systems*, vol.**3** (2002).

[8] H. K Khalil. Extended High-Gain Observers as Disturbance Estimators, SICE Journal of Control, Measurement and System Integration, **Vol. 10**, **No. 3**, pp. 125–134, May (2017)

[9] H. K Khalil, L. Praly, Generalized Extended State Observer Based High Precision Attitude Control of Quadrotor Vehicles Subject to Wind Disturbance, International Journal of Robust and Nonlinear Control, **24**, 993–1015, (2014).

[10] DI Shi, Zhong Wu AND Wusheng Chou, *High-gain observers in nonlinear feedback control*, IEEE Access, **Volume 6**, (2018).

A Fault Inference Method under Uncertainty: Case Study on Crankshafts in Fracturing Pumps

X Zhang[1]**, L B Zhang**[1] **and J Q Hu**[1]

[1] College of Safety and Ocean Engineering, China University of Petroleum, 102249 Beijing, China

E-mail: hujinqiu@cup.edu.cn

Abstract. Crankshaft is a pivotal mechanical unit in the power-end system of a fracturing pump and its fault inference could facilitate optimal condition-based maintenance. Fracturing pumps are equipped with advanced instrumentation systems able to acquire vibration information for crankshaft fault analysis, but there exist complex uncertain dependences between faults and symptoms as well as incomplete symptom information, further increasing the difficulty of fault inference by operators. To achieve effective fault inference in the case of uncertain or incomplete diagnosis evidences, a Bayesian network-based fault inference method for crankshafts is proposed in this article. The approach can be utilized to implement cause inference and diagnosis inference by incorporating cause nodes, fault nodes and symptom nodes into a Bayesian network (BN) model. The application of the presented approach in fault inference of crankshafts indicates its strong inference capability under uncertainty. The results from the presented BN model may offer a useful aid to repairers in their maintenance decision-making processes.

1. Introduction

In recent years, shale gas as a clean energy has attracted increasing attention over the world. To enable efficient development of shale-gas wells, large-scale fracturing pumps have been extensively utilized for shale-gas wells fracturing operations [1]. These pumps could pressurize the sand-laden fluid and then pump it into the high-pressure manifolds. Owing to the harsh working conditions such as high-pressure, large-displacement and super-power, mechanical units or components suffer from rapid degradation. In particular, the crankshaft is always subjected to rapid deterioration due to its high-speed and high-load. The enhanced equipment safety management has revealed a need for a more accurate and more intelligent fault inference method. Machinery fault inference is an inverse reasoning process from symptoms to faults. In engineering, there are multiple corresponding relations between symptoms and faults, including one-to-one, one-to-many and many-to-one correspondence [2]. Moreover, the collected symptom information is sometimes incomplete or uncertain because of the limited data acquisition conditions. Therefore, it is essential to carry out fault inference under conditions of uncertainty or incompleteness. This situation is especially suitable for the crankshafts because of the complex correspondences between faults and symptoms as well as the difficulties in collecting and analyzing crankshaft's vibration signals.

Fault diagnosis of equipment or components has received considerable attention over the last decade [3–11]. Various artificial intelligence methods have been developed for fault diagnosis, including support vector machine [4,5,10], neural network [6–9], fuzzy logic [3], and their hybrid methods. For example, Ref. [10] have developed an intelligent fault diagnosis technique for locomotive roller bearings by combining support vector machine and ant colony algorithm; in this technique, the ant colony algorithm was used for synchronous feature selection and parameter optimization for support vector machine. Ref. [9] have adopted artificial neural networks to isolate and identify eight different types of faults in photovoltaic systems. In this method, features were manually extracted from a number of simulations. Ref. [11] have proposed a promising tool on a basis of deep neural networks for fault feature mining and intelligent diagnosis of rotating machinery with massive data.

Although the above methods have significantly promoted the progress of fault analysis of mechanical units, they usually depend on complete signal data and thus have inadequate robustness in the presence of incomplete or uncertain diagnosis data. Hence, a novel intelligent fault inference method for crankshafts is highly required. Motivated by above requirement, this paper proposes an intelligent fault inference approach under uncertainty based on Bayesian network (BN) by taking full advantage of expert knowledge, fault causes, faults and symptoms. The proposed approach fully takes into account the complex uncertainties between causes, faults and symptoms, as well as the incomplete symptoms. Using this method, the cause inference and diagnosis inference could be implemented.

The remainder of this article is structured as follows: Section 2 gives a brief overview of BN. Section 3 describes the proposed method for developing an intelligent fault inference model for crankshafts. In Section 4, the proposed approach is applied for fault inference of a crankshaft in a fracturing pump. Finally, the conclusions are given in Section 5.

2. Bayesian network

A Bayesian network as a directed acyclic graph (DAG) is suitable for a visual modelling of the causality with uncertainty. The analysis of a Bayesian network is based on the specification of conditional probabilities of child nodes for given states of their parent nodes, utilizing the concept of conditional probabilities table (CPT). The nodes in a BN can embrace multiple states, and the directed arrows mean the probabilistic dependencies among nodes. For more general information on BN, see Nielsen and Jensen. [12].

Bayesian networks have proved to be a powerful tool in various engineering areas; recent applications from a range of technical fields include reliability evaluation of marine floating structures [13] and kick control operation [14], risk analysis of marine transportation [15] and submarine pipelines [16], forecast of stock market index [17] and forensic assessments [18]. These studies have been motivated by the strong points of Bayesian networks that facilitate:

(1) Modelling the complex causal interdependence in a graphical way, with clear semantics and an easy-to-understand structure.

(2) Updating the probability distribution of target nodes when new evidences are available.

3. Proposed methodology

The steps for the proposed methodology is illustrated in figure 1. The steps are stated in detail as follows:

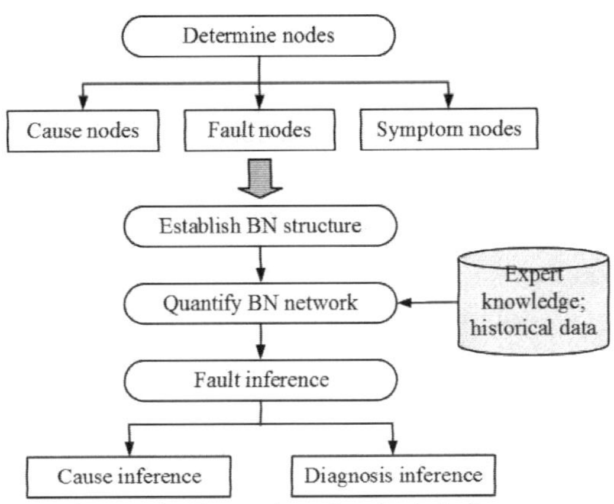

Figure 1. Procedure for fault inference.

3.1. Step 1: Determine nodes
The built BN model in this paper is composed of three classes of nodes: cause nodes (root nodes), fault nodes (intermediate nodes) and symptom nodes (leaf nodes). A cause is defined as an internal or external environmental factor of abnormal conditions that could lead to the loss of ability to perform a required function.

For example, poor thermal stability is the cause of crankshaft imbalance. A fault is a condition that renders an element unable to perform its required function at the expected levels of performance. A symptom may be defined as a deviation from the expected behaviour of a system or a specific phenomenon sensitive to the concerned faults. For instance, the symptoms of crankshaft misalignment may include strong axial vibration and doubling-frequency. These nodes could be determined from failure mode and effect analysis (FMEA), or fault tree analysis (FTA) of machinery or from the equipment maintenance staff.

3.2. Step 2: Establish BN structure
In fact, when various nodes are defined in Step 1, the complex uncertain relations among these nodes are also analyzed and determined. The purpose of this step is to establish the BN structure according to their complex inter-relationships. The constructed BN network contains two topological configurations of causality: topology structure between cause nodes and fault nodes as well as that between fault nodes and symptom nodes.

3.3. Step 3: Quantify BN network
The parameters in a BN consist of prior probabilities of cause nodes and conditional probabilities amongst nodes. Prior probabilities can be assigned by analyzing historical fault reports or by expert knowledge. Conditional probabilities could be estimated using parameter learning methods such as Maximum Likelihood Estimation (MLE) or expert knowledge. For more details on the methods to estimate the conditional probabilities, please refer to Reference [12].

3.4. Step 4: Fault inference
Fault inference can be further divided into diagnosis inference and cause inference. Diagnosis inference is a process of updating prior probabilities of fault nodes using subsequent available information from cause nodes or symptom nodes. Based on the observed evidence, the occurrence probabilities of each fault could be obtained using the following Bayesian theorem:

$$P\left(F_i|e\right)=\frac{P\left(F_i e\right)}{P(e)}=\frac{P\left(e|F_i\right)P\left(F_i\right)}{\sum_{i=1}^{n}P\left(e|F_i\right)P\left(F_i\right)} \tag{1}$$

in which $P\left(e|F_i\right)$ is the conditional probability of evidence e given the fault F_i; $P\left(F_i\right)$ is the prior probability of F_i; $P(e)=\sum_{i=1}^{n}P\left(e|F_i\right)P\left(F_i\right)$ is the marginal probability of evidence; $P\left(F_i|e\right)$ is called the posterior probability given the evidences e.

Then the most probable fault can be identified. Causal inference is to estimate posterior probabilities of ancestor nodes of the observed evidences. Causal inference is applicable to computing the magnitude of a cause effect on faults, or identify the likely causes after updating the symptoms.

4. Case study
In this section, the proposed BN method is applied to fault inference of crankshafts. Crankshafts play a significant role in ensuring the smooth operation of the pumps. In the construction site, multiple fault modes such as imbalance, misalignment and bending of crankshaft could result in performance degradation of pumps and even undesirable events. These faults may have some identical fault symptoms; moreover, one type of fault with varying degrees may have different symptoms, indicating a complex uncertainty relations between the faults and the symptoms. Therefore, fault inference for crankshafts has become a challenging but also intellectually fascinating task in engineering.

4.1. Step 1: Determine nodes

In this case, five main fault modes of a crankshaft are selected as fault nodes, as shown in Table 1. Through field investigation and literature research, ten cause nodes and twelve symptom nodes are identified and listed in table 1. The symptom nodes can be roughly categorized as four classes: vibration direction nodes (S1 and S2), axis orbit nodes (S3, S4 and S5), characteristic frequency nodes (S6, S7, S8 and S9) as well as secondary frequency nodes (S10, S11 and S12).

Table 1. Information of nodes.

No.	Node	Description	No.	Node	Description
1	C1	Initial bending	15	F5	Crankshaft cracks
2	C2	Misalignment of bearing block	16	S1	Abnormal axial vibration
3	C3	Uneven material	17	S2	Abnormal radial vibration
4	C4	Poor thermal stability	18	S3	Axis orbit with single ellipse
5	C5	Polluted lubricant	19	S4	Axis orbit with double ellipse
6	C6	Unreasonable axle bush	20	S5	Irregular axis orbit
7	C7	Excessive lubricant temperature	21	S6	0.5X characteristic frequency
8	C8	Excessive load	22	S7	1X characteristic frequency
9	C9	Frequent starting	23	S8	2X characteristic frequency
10	C10	Fatigue running	24	S9	3X characteristic frequency
11	F1	Crankshaft Misalignment	25	S10	1X secondary frequency
12	F2	Crankshaft bending	26	S11	2X secondary frequency
13	F3	Crankshaft imbalance	27	S12	Higher harmonics
14	F4	Oil-film whirl			

4.2. Step 2: Establish BN structure

By analysing the uncertain causal relations among the identified nodes, the BN network is built and provided in figure 2. As observed in figure 2, there exist complex dependencies among nodes. For instance, uneven material may lead to three faults at the same time and abnormal axial vibration may be caused by five faults.

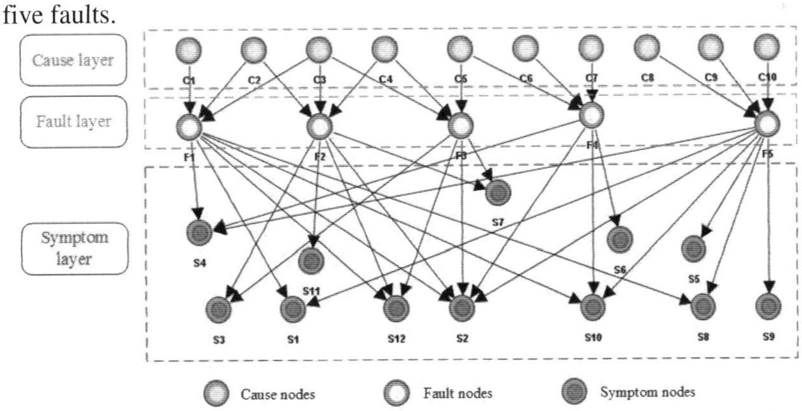

Figure 2. Bayesian network for crankshafts.

4.3. Step 3: Quantify BN network

Quantifying BN structure is a challenging issue because of the complicated causality. Limited fault and maintenance records could be accessed due to their confidentiality; thus, the prior probabilities of cause nodes are assigned by means of expert knowledge and given in table 2. The conditional probabilities between cause nodes and fault nodes are also determined using expert judgment. For basic sensing information relevant to symptoms such as axis orbit, advanced instrumentation system in pumps provide an access to them; so sensing information from February 2018 to June 2018 is collected and utilized to learn conditional probabilities between fault nodes and symptom nodes. Appendix provides a part of dataset used in this case study. Taking four nodes (F1, F4, F5 and S4) as an example, the CPT among them is given in table 3.

Table 2. Prior probability of cause nodes.

Node	State	Prior probability	Node	State	Prior probability
C_1	Yes	3%	C_6	Yes	7%
	No	97%		No	93%
C_2	Yes	5%	C_7	Yes	15%
	No	95%		No	85%
C_3	Yes	8%	C_8	Yes	10%
	No	92%		No	90%
C_4	Yes	7%	C_9	Yes	18%
	No	93%		No	82%
C_5	Yes	13%	C_10	Yes	15%
	No	87%		No	85%

Table 3. Conditional probabilities among four nodes.

Fault			S4 (%)	
F1	F4	F5	Yes	No
Yes	Yes	Yes	100	0
Yes	Yes	No	100	0
Yes	No	Yes	100	0
Yes	No	No	97.27	2.73
No	Yes	Yes	99.46	0.54
No	Yes	No	94.35	5.65
No	No	Yes	96.89	3.11
No	No	No	4.62	95.38

4.4. Step 4: Fault inference

Before conducting a case study, the concepts of hard evidences and soft evidences must be defined. Hard evidences refer to certain symptoms whose probabilities of occurrence are 100%, while soft evidences are those whose probabilities are between 0 and 100%. Next, a case with simple evidences and two cases with complex evidences will be provided. Simple evidences is a combination of abnormal symptoms in which the class number of hard evidences is no more than three, while complex evidences refers to the combination of abnormal symptoms in which the class number of hard evidences is more than three. Worth noting that hard evidences in the complex symptoms may come from the cause layer or the symptom layer.

4.4.1. Case 1: fault inference under simple evidences. In the first case, only two hard evidences appear: abnormal radial vibration (S2 is yes) and obvious 1X characteristic frequency in the frequency spectrum (S7 is yes). The probability of single-ring ellipse (S3) is set to be 30%. Obviously, it can be seen qualitatively from figure 2 that crankshaft bending (F2) or crankshaft imbalance (F3) are likely to occur. The above symptoms are input into the BN model to update prior probabilities. The obtained posterior probability of each fault node are listed in table 4. The posterior probability of F3 (42.87%) is the largest, followed by F2 (1.38%), indicating the most likely fault is crankshaft imbalance. In the cause layer, the probability of polluted lubricant oil C5 (16.68%) is maximum, showing that C5 is the cause of F3. In order to reduce the damage to fracturing pump caused by impurities in lubricating oil, it is necessary to replace the oil filter or to replace the polluted lubricant in time

Table 4. Information of fault inference.

No.	Evidence		Fault inference		
	Complexity	Description	Probability of fault	Diagnosis inference	Cause inference
Case 1	Simple (symptom layer)	S2 is yes S7 is yes P(S3) = 30%	*P*(F1|e) = 3.54% *P*(F2|e) = 11.38% *P*(F3|e) = 42.87% *P*(F4|e) = 1.86% *P*(F5|e) = 0.27%	F3 is yes.	C5 is yes. *P*(C5|e) = 16.68%.
Case 2	Complex (symptom layer)	S2 is yes S4 is yes S8 is yes S10 is yes P(S1) = 20%	*P*(F1|e) = 50.04% *P*(F2|e) = 3.89% *P*(F3|e) = 6.74% *P*(F4|e) = 45.50% *P*(F5|e) = 0.73%	F1 and F4 are yes.	C7 and C3 are yes. *P*(C7|e) = 19.82% and P(C3|e) = 9.73%.
Case 3	Complex (symptom layer and cause layer)	C10 is yes S1 is yes S2 is yes S5 is yes S8, S10 are yes	*P*(F1|e) = 52.05% *P*(F2|e) = 16.47% *P*(F3|e) = 4.52% *P*(F4|e) = 3.27% *P*(F5|e) = 46.78%	F1 and F5 are yes.	C3 is yes. *P*(C3|e) = 31.42%.

4.4.2. Case 2: fault inference under complex evidences. In the second case, there are four hard evidences and one soft evidence from the symptom layer: abnormal radial vibration (S2 is yes), axis orbit with double ellipse (S4 is yes), and obvious 2X characteristic frequency (S8 is yes) along with 1X secondary frequency (S10 is yes). Because the axial vibration is slightly worse than that in normal condition, the occurrence probability of S1 is set to be 20%. The inference results are listed in table 4. As shown in table 4, the crankshaft misalignment F1 and the oil-film whirl F4 have the similar posterior probabilities (50.04% and 45.50% respectively), indicating that there is a composite fault of F1 and F4. In the cause layer, excessive lubricating oil temperature C7 and uneven material C3 are the most likely causes of the above faults.

4.4.3. Case 3: fault inference under complex evidences. In the third case, the observed evidences come from cause layer and symptom layer. The hard evidences include fatigue running of crankshaft (C10 is yes), irregular axis orbit (S5 is yes), abnormal radial vibration and axial vibration (S1 is yes and S2 is yes), and obvious 2X characteristic frequency (S8 is yes) along with 1X secondary frequency (S10 is yes). The structure of BN denotes that fatigue state of crankshaft C10 would give rise to cracks on its body. However, the inference results show that the crankshaft misalignment F1 and crankshaft cracks F5 are likely to exist at the same time because of their similar posterior probabilities (52.05% and 46.78%). Moreover, the cause inference indicates F1 is caused by uneven material of crankshaft C3.

5. Conclusion

Fault inference of crankshafts in fracturing pumps is a challenging but also intellectually appealing task in engineering. Although fracturing pumps are installed with data acquisition systems that could acquire sensing information for fault analysis, there exist complex uncertain relations between faults and symptoms as well as incomplete symptom information, further increasing the difficulty of fault inference by human.

To perform fault inference under conditions of uncertainty and incompleteness, this paper proposes an intelligent fault inference method for crankshafts using a Bayesian network. The presented method integrates cause nodes, fault nodes and symptom nodes into a BN model. Using forward-backward reasoning, fault inference and cause inference can be achieved. The application of the presented approach in fault inference of crankshafts indicates its strong inference ability under uncertain and incomplete information. The obtained results can provide scientific information for equipment maintenance support decision. Further research is required to develop real-time diagnosis system based on the proposed method and to broaden its application range.

Acknowledgements
This work was supported by the National Natural Science Foundation of China [Grant numbers: 51574263], and Science Foundation of China University of Petroleum-Beijing [Grant numbers: 2462015YQ0403, Grant numbers: C201602]. The BN model in this paper was created using BNT toolbox in MATLAB developed by Kevin Murphy. The authors would like to thank the engineers in the field for their valuable inputs to this article.

Appendix

Table A1: Partial dataset used in this study

No.	F1	F2	F3	F4	F5	S1	S2	S3	S4	S5	S6	S7	S8	S9	S10	S11	S12
1	No	No	No	No	No	No	No	No	No	No	No	No	No	No	No	No	No
2	No	No	No	No	No	No	No	No	No	No	No	No	No	No	No	No	No
3	No	No	No	No	No	No	No	No	No	No	No	No	No	No	No	No	No
…	…	…	…	…	…	…	…	…	…	…	…	…	…	…	…	…	…
105	Yes	No	No	No	No	Yes	No	No	Yes	No	No	No	Yes	No	Yes	No	No
106	Yes	No	No	No	No	Yes	No	No	Yes	No	No	No	Yes	No	Yes	No	No
107	Yes	No	No	No	No	Yes	No	No	Yes	No	No	No	Yes	No	Yes	No	Yes
…	…	…	…	…	…	…	…	…	…	…	…	…	…	…	…	…	…
976	No	No	Yes	No	No	No	No	Yes	No	No	No	Yes	No	No	No	No	No
977	No	No	Yes	No	No	No	Yes	Yes	No	No	No	Yes	No	No	No	No	No
978	No	No	Yes	No	No	No	Yes	Yes	No	No	No	Yes	Yes	No	No	No	Yes
…	…	…	…	…	…	…	…	…	…	…	…	…	…	…	…	…	…
4836	No	Yes	No	No	No	No	No	Yes	No	No	No	Yes	No	No	No	Yes	Yes
4837	No	Yes	No	No	No	No	Yes	Yes	No	No	No	Yes	No	No	No	Yes	Yes
4838	No	Yes	No	No	No	Yes	Yes	Yes	No	No	No	Yes	No	No	No	Yes	Yes
…	…	…	…	…	…	…	…	…	…	…	…	…	…	…	…	…	…
9580	No	No	No	Yes	No	No	Yes	No	Yes	No	Yes	No	No	No	Yes	No	No
9581	No	No	Yes	Yes	No	No	Yes	No	Yes	No	Yes	No	No	No	Yes	No	No
9582	No	No	Yes	Yes	No	No	Yes	No	No	Yes	Yes	Yes	No	No	Yes	No	No
…	…	…	…	…	…	…	…	…	…	…	…		…	…	…	…	…

References

[1] Zhang X, Zhang L and Hu J 2018 Real-time diagnosis and alarm of down-hole incidents in the shale-gas well fracturing process *Process Saf. Environ. Prot.* **116** 243–53
[2] Hu J, Zhang L, Ma L and Liang W 2010 An integrated method for safety pre-warning of complex system *Saf. Sci.* **48** 580–97
[3] Simani S, Farsoni S and Castaldi P 2015 Wind turbine simulator fault diagnosis via fuzzy modelling and identification techniques *Sustain. Energy, Grids Networks* **1** 45–52
[4] Ayodeji A and Liu Y kuo 2018 Support vector ensemble for incipient fault diagnosis in nuclear plant components *Nucl. Eng. Technol.* 1–8
[5] Yang D, Liu Y, Li S, Li X and Ma L 2015 Gear fault diagnosis based on support vector machine optimized by artificial bee colony algorithm *Mech. Mach. Theory* **90** 219–29
[6] Huang S R, Huang K H, Chao K H and Chiang W T 2016 Fault analysis and diagnosis system for induction motors *Comput. Electr. Eng.* **54** 195–209

[7] Janssens O, Slavkovikj V, Vervisch B, Stockman K, Loccufier M, Verstockt S, Van de Walle R and Van Hoecke S 2016 Convolutional Neural Network Based Fault Detection for Rotating Machinery *J. Sound Vib.* **377** 331–45

[8] Merabet H, Bahi T and Halem N 2015 Condition Monitoring and Fault Detection in Wind Turbine Based on DFIG by the Fuzzy Logic *Energy Procedia* **74** 518–28

[9] Chine W, Mellit A, Lughi V, Malek A, Sulligoi G and Massi Pavan A 2016 A novel fault diagnosis technique for photovoltaic systems based on artificial neural networks *Renew. Energy* **90** 501–12

[10] Zhang X L, Chen W, Wang B J and Chen X F 2015 Intelligent fault diagnosis of rotating machinery using support vector machine with ant colony algorithm for synchronous feature selection and parameter optimization *Neurocomputing* **167** 260–79

[11] Jia F, Lei Y, Lin J, Zhou X and Lu N 2016 Deep neural networks: A promising tool for fault characteristic mining and intelligent diagnosis of rotating machinery with massive data *Mech. Syst. Signal Process.* **72–73** 303–15

[12] Nielsen T D and Jensen F V 2009 *Bayesian Network and Decision Graph*

[13] Abaei M M, Abbassi R, Garaniya V, Chai S and Khan F 2018 Reliability assessment of marine floating structures using Bayesian network *Appl. Ocean Res.* **76** 51–60

[14] Sule I, Khan F, Butt S and Yang M 2018 Kick control reliability analysis of managed pressure drilling operation *J. Loss Prev. Process Ind.* **52** 7–20

[15] Baksh A A, Abbassi R, Garaniya V and Khan F 2018 Marine transportation risk assessment using Bayesian Network: Application to Arctic waters *Ocean Eng.* **159** 422–36

[16] Li X, Chen G and Zhu H 2016 Quantitative risk analysis on leakage failure of submarine oil and gas pipelines using Bayesian network *Process Saf. Environ. Prot.* **103** 163–73

[17] Malagrino L S, Roman N T and Monteiro A M 2018 Forecasting stock market index daily direction: A Bayesian Network approach *Expert Syst. Appl.* **105** 11–22

[18] Holický M, Marková J and Sýkora M 2013 Forensic assessment of a bridge downfall using Bayesian networks *Eng. Fail. Anal.* **30** 1–9

IOP Conf. Series: Materials Science and Engineering **575** (2019) 012011 doi:10.1088/1757-899X/575/1/012011

A Design Method of Two DOF Joint at the Base of Heavy-load Manipulator Based on Linear Driving

Zongmiao Dai[1], Peizhang Wu[2] and Qingyuan Wang[2]

[1]Mechanical Department, the 713th Institute of China Shipbuilding Industry Corporation, Zhengzhou450015, Henan Province, China
[2]Automation Department, the 713th Institute of China Shipbuilding Industry Corporation, Zhengzhou450015, Henan Province,China

daizongmiao@tom.com

Abstract. In this paper, we propose a two DOF joint mechanical design scheme for heavy-load manipulator based on linear driving, the key structural parameters that directly influencing on the linear driving force is determined. Based on the multi-body kinematics and the principle of virtual work, we derive the relationship between the torque of joint space at the base of the heavy-load manipulator and the force of linear driving, then by analyzing the relationship, we obtain the design method of key structural parameters. Finally, using simulation analysis we validate this design method.

1. Introduction

Since 1913, the Ford Motor Co. has been using automatic production line instead of manual production for more than 100 years[1-6]. Automatic production line has been developed rapidly in America, Europe, Japan and other developed countries. the most widely used in automatic production is the serial multi-joint manipulator, which is usually set with two rotational joint at the base. The two rotational joint consist of pitch rotation and yaw rotation, which are usually drived by motor and gear reducer.

In order to meet the requirement of larger workspace, the mechanical arms are often longer. Moreover, it needs a very large pitch torque to act on the base of mechanical for lifting a heavy loader. Thus the base of manipulator structure is usually designed bulky, as well as the ratio of payload-weight to self-weight is very low. In view of mentioned above, we propose a design scheme of two DOF joint at the base of heavy-load manipulator based on linear driving. It has the advantage of large output torque but with light self-weight, which sharply increase the ratio of payload-weight to self-weight.

2. Design Scheme

The design scheme for the two DOF joint at base of heavy-load manipulator based on linear driving is mainly composed of main arm yaw joint(Fig.1-(1)), main arm pitch joint(Fig.1-(2)), main arm(Fig.1-(3)), main arm left/right end joint(Fig.1-(4/5)), left/right link (Fig.1-(6/7)), left/right slider end joint(Fig.1-(8/9)), left/right slider(Fig.1-(10/11)), left/right slideway(Fig.1-(12/13)), left/right linear driving mechanism(Fig.1-(14/15)), etc. In Figure 1, the main arm's yaw/pitch joint is the single DOF joint, the main arm's left/right end joint and the left/right slider's end joint both are three DOF joint, the left/right linear driving mechanism can be driven by hydraulic cylinder, electric cylinder or nut screw. When driving the left (right) slider of linear motion along the left (right) slideway, the yaw and pitch rotating of the main arm can be realized.

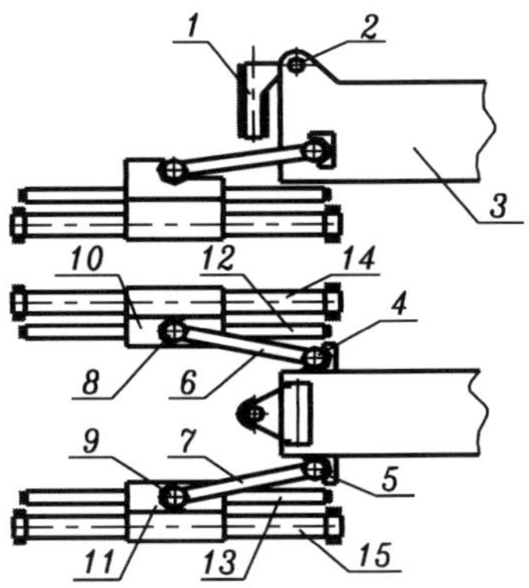

Figure 1. Composition of base structure

3. Design Method of Key Structure Parameters

3.1. Analysis of kinematics

In this design scheme, the key structural parameters of yaw and pitch torque are determined by the following parameters: the common perpendicular distance a1 is between the main arm yaw joint and pitch joint, the horizontal distance d2 is between the main arm yaw joint and left/right end joint, the vertical distance a2 is between the main arm pitch joint and left/right end joint, the L is the length of left/right link, the horizontal distance a3 is between the main arm yaw joint and left/right slider end joint, the vertical distance d1 is between the main arm pitch joint and left/right slider end joint (see Fig. 2).

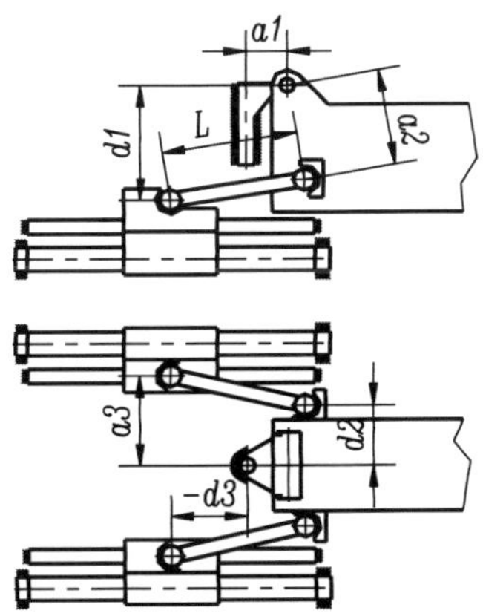

Figure 2. Key parameters of base basestructure

A coordinate system for the two DOF joint at the base of heavy-load manipulator based on linear driving is established by analyzing multi-body kinematics. In Fig.3, the O0 is the origin of base coordinate frame, the O1 is the origin of coordinate frame for main arm yaw joint, Z1 is the rotation axis of yaw joint, O2 is the origin of coordinate frame for main arm pitch joint, Z2 is the rotation axis of pitch joint, O3 is the origin of coordinate frame for main arm right end joint.

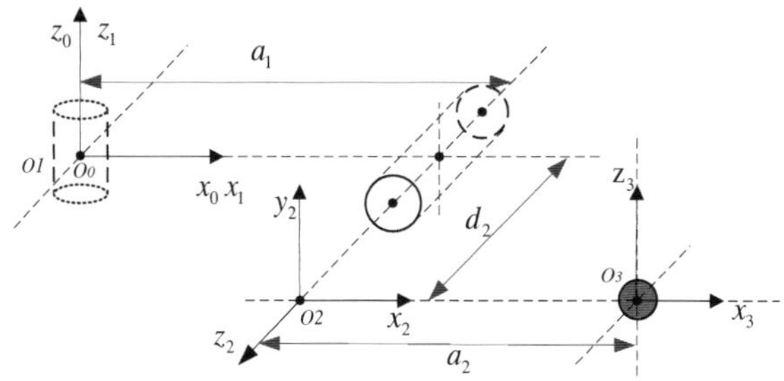

Figure 3. Composition of base structure

3.2. Principle of virtual work

According to reference[7], we use homogeneous coordinate transformation method to obtain the motion trajectory equation of the main arm left/right end joint.

$$\begin{cases} x_L = a_1 cos\theta_1 - d_2 sin\theta_1 + a_2 cos\theta_1 cos\theta_2 \\ y_L = a_1 sin\theta_1 + d_2 cos\theta_1 + a_2 sin\theta_1 cos\theta_2 \\ z_L = a_2 sin\theta_2 \end{cases} \tag{1}$$

$$\begin{cases} x_R = a_1 cos\theta_1 + d_2 sin\theta_1 + a_2 cos\theta_1 cos\theta_2 \\ y_R = a_1 sin\theta_1 - d_2 cos\theta_1 + a_2 sin\theta_1 cos\theta_2 \\ z_L = a_2 sin\theta_2 \end{cases} \tag{2}$$

In Formula (1) and (2), X_L/X_R, Y_L/Y_R, Z_L/Z_R, isthe *X/Y/Z* position of main arm left/right end joint in the base coordinate frame, respectively.$\theta_1$is the angle of main arm yaw rotation, $\theta_2$is the angle of main arm pitch rotation.

Given that the designing feasibility for the joint structure layout, we assumption that-$90°\le\theta_1\le 90°$, $0°\le\theta_2\le 90°$. In Fig. 2, we can get the equation from the geometric relationships as following:

$$L^2=(d_{3L}-X_L)^2+(a_3-y_L)^2+(d_1+z_L)^2 \tag{3}$$

$$L^2=(d_{3R}-X_R)^2+(a_3+y_R)^2+(d_1+z_L)^2 \tag{4}$$

In Formula (3) and (4), d3L/d3R is respectively the displacement of left/right slider end joint moving along left/right slideway.

Substituting the Formula (1) into (3) and Formula (2) into (4), we can get the relationships as following:

$$d_{3L}=-A+ a_1cos\theta_1- d_2\ sin\theta_1+a_2cos\theta_1cos\theta_2 \tag{5}$$

$$d_{3R}=-B+ a_1cos\theta_1+ d_2\ sin\theta_1+a_2cos\theta_1cos\theta_2 \tag{6}$$

In Formula (5) and (6),

$$A = \sqrt{\begin{matrix} L^2 - (a_3 - a_1 sin\theta_1 - d_2 cos\theta_1 - a_2 sin\theta_1 cos\theta_2)^2 \\ -(d_1 + a_2 sin\theta_2)^2 \end{matrix}}$$

$$B = \sqrt{\begin{matrix} L^2 - (-a_3 - a_1 sin\theta_1 + d_2 cos\theta_1 - a_2 sin\theta_1 cos\theta_2)^2 \\ -(d_1 + a_2 sin\theta_2)^2 \end{matrix}}$$

When $\theta_1 = 0°, \theta_2 = 0°, d_{3L} = d_{3R}$,from Formula (5) and (6), we can get the relationships as following:

$$L^2=(d_{3L}- a_1-a_2)^2+(a_3-d_2)^2+d_1^2 \tag{7}$$

When $\theta_1 = 0°, \theta_2 = -90°, d_{3L} = d_{3R}$, from Formula (5) and (6), we can get therelationships as following:

$$L^2=(d_{3L}- a_1)^2+(a_3-d_2)^2+(d_1+a_2)^2 \tag{8}$$

When $\theta_1 = 90°, \theta_2 = 0°, d_{3L} = d_{3R}$, from Formula (5) and (6), we can get therelationships as following:

$$L^2=(d_{3L}+ d_2)^2+(a_3- a_1-a_2)^2+d_1^2 \tag{9}$$

$$L^2= (d_{3R}- d_2)^2 + (a_3+ a_1+a_2)^2+d_1^2 \tag{10}$$

When$\theta_1 = 90°, \theta_2 = -90°, d_{3L} = d_{3R}$, from Formula (5) and (6), we can get therelationships as following:

$$L^2 = (d_{3L}+ d_2)^2+(a_3- a_1)^2+(d_1+a_2)^2 \tag{11}$$

$$L^2 = (d_{3R}- d_2)^2+(a_3+ a_1)^2+(d_1+a)^2 \tag{12}$$

From Formula (7) to (12), when $(a_3 + a_3) \geq d_1$, to ensure the base yaw joint angle can reach $-90° \leq \theta_1 \leq 90°$, and the base pitch joint angle can reach $0° \leq \theta_2 \leq 90°$, the length L of left/right link must meet the following requirements:

$$L^2 \geq (a_3+ a_1+a_2)^2+d_1^2 \tag{13}$$

In order to make the manipulator more compact and lighter, we can set L as following:

$$L= \sqrt{(a_3 + a_1 + a_2)^2 + {d_1}^2} \tag{14}$$

According to the reference[8], we use the principle of virtual work. Under some assumptions, there are two linear forceF_L and F_R respectively act on the left and right slider end joint, which can cause two small differential displacement δd_{3L} and δd_{3R}. The virtual work is as following:

$$\partial W = \boldsymbol{F}^T \partial \boldsymbol{d}_3 \tag{15}$$

In Formula (15), $\boldsymbol{F}^T=[F_L\ F_R]$, $\partial \boldsymbol{d}_3=[\partial d_{3L}, \partial d_{3R}]^T$.

At the same time, it also causes the output torque T_1 and T_2 with rotating small differential angle θ_1 and θ_2 of main arm yaw and pitch joint, respectively. The virtual work is:

$$\partial W = \boldsymbol{T}^T \partial \boldsymbol{\theta} \tag{16}$$

In Formula (16), $\boldsymbol{T}^T= [T_1\ F_R]$, $\partial \boldsymbol{\theta}= [\partial\theta_1, \partial\theta_2]^T$.

From principle of virtual work, the Formula (15) equals (16) as following:

$$\boldsymbol{F}^T \partial \boldsymbol{d}_3 = \boldsymbol{T}^T \partial \boldsymbol{\theta} \tag{17}$$

From Formula (5) and (6), we can simply express the relationship $\boldsymbol{d}_3=\boldsymbol{d}_3(\boldsymbol{\theta})$, then we can establish the velocity relationship:

$$\dot{\boldsymbol{d}}_3 = \boldsymbol{J}(\boldsymbol{\theta})\dot{\boldsymbol{\theta}} \tag{18}$$

In Formula (18), $\dot{d}_3 = [\dot{d}_{3L}, \dot{d}_{3R}]^T$, $\dot{\theta} = [\dot{\theta}_1, \dot{\theta}_2]^T$,$\boldsymbol{J}(\boldsymbol{\theta})$ is the velocity Jacobian matrix as following:

$$J_{ij}(\theta) = \frac{\partial d_{3i}(\theta)}{\partial \theta_j} \tag{19}$$

From Formula (18), we can get:

$$\partial \boldsymbol{d}_3 = \boldsymbol{J}(\boldsymbol{\theta}) \partial \boldsymbol{\theta} \tag{20}$$

Solving the simultaneous equations (17) and (20), we can get:

$$\boldsymbol{T} = \boldsymbol{J}(\boldsymbol{\theta})^T \boldsymbol{F} \tag{21}$$

From Formula (19) and (21), we can get:

$$\begin{cases} T_1 = \frac{\partial d_{3L}}{\partial \theta_1} F_L + \frac{\partial d_{3R}}{\partial \theta_1} F_R \\ T_2 = \frac{\partial d_{3L}}{\partial \theta_2} F_L + \frac{\partial d_{3R}}{\partial \theta_2} F_R \end{cases} \tag{22}$$

In Formula (22), T_1 is the yaw torque of main arm, which is used for overcoming the friction and rotation inertia moment, and T_1 is much smaller than T_2 when the yaw rotation acceleration is low. T_2 is the pitch torque of main arm, which is used for lifting heavy load, and it importantly measures the load capacity of the manipulator. So we emphatically analyze the key structure parameters that how to influence on the T_2. For simplicity, we make $\theta_1 = 0°, 90°$and $\theta_2 = 0°, 90°$to formulate the base pitch torque of main arm at four limit positions.

3.3. Analysis of parameters

When$\theta_1 = 0°, \theta_2 = 0°$, substituting $F_L = F_R = F_1$ into Formula (21), we can get the base pitch torque T2 of main arm:

$$T_{21} = \frac{2d_1F_1}{\sqrt{e^2+2e+2eg+(2fg-f^2)+2g+1}} \tag{23}$$

In Formula (23), $e = a_1/a_2$, $f = d_2/a_2$, $g = a_3/a_2$. Given the same input force F_1, in order to output a higher torque T_2, we should make d_1 as large as possible, and make e, g, $2fg$-f^2 assmall as possible. So the parameters should as following:

$$\begin{cases} a_2 \gg a_1 \\ a_2 \gg a_3 \\ a_3 = d_2 \\ d_1 \gg 0 \end{cases} \tag{24}$$

When $\theta_1 = 0°, \theta_2 = 90°$, substituting $F_L = F_R = F_2$ into Formula (21), we can get the base pitch torque T_2 of main arm:

$$T_{22} = -2a_2F_2 \tag{25}$$

Given the same input force F_2, in order to output a higher torque T_2, we should make a_2 as large as possible. So the parameters should as following:

$$a_2 \gg 0 \tag{26}$$

When $\theta_1 = 90°, \theta_2 = 0°$, from Formula (21), we can get the base pitch torque T_2 of main arm:

$$T_{23} = \frac{d_1a_2}{\sqrt{L^2-(a_3-a_1-a_2)^2-d_1^{\,2}}} F_L + \frac{d_1a_2}{\sqrt{L^2-(a_3+a_1+a_2)^2-d_1^{\,2}}} F_R \tag{27}$$

Substituting Formula (14) into Formula (27), we can get:

$$\frac{d_1a_2}{\sqrt{L^2-(a_3-a_1-a_2)^2-d_1^{\,2}}} F_L = \frac{d_1F_L}{\sqrt{4eg+4g}} \tag{28}$$

The right second expression of Formula (27) is infinite, so it is a singularity position for this manipulator.

When $\theta_1 = 90°, \theta_2 = 90°$, from Formula (21), we can get the base pitch torque T_2 of main arm:

$$\mathrm{T}_{24} = \frac{\mathrm{a}_3-\mathrm{a}_1}{\sqrt{4\mathrm{eg}+2\mathrm{e}+2\mathrm{g}-2\mathrm{w}}} \mathrm{F}_\mathrm{L} + \frac{-(\mathrm{a}_1+\mathrm{a}_3)}{\sqrt{2\mathrm{e}+2\mathrm{g}-2\mathrm{w}}} \mathrm{F}_\mathrm{R} \tag{29}$$

If $F_L < 0$ *and* given the same input force F_L *and* F_R, in order to output a higher torque T_2, we should make parameters as following:

$$\begin{cases} a_2 \gg a_1 \\ a_2 \gg a_3 \\ a_3 \gg a_1 \\ d_1 \leq a_1 + a_3 \end{cases} \tag{30}$$

Taking into consideration of Formula (24), (26) and (30), when the manipulator working in the range -90° ≤ θ_1 ≤ 90°, 0°≤ θ_2 ≤ 90, if getting more driving torque, it should be constrained by the following parameters:

$$\begin{cases} a_2 \gg a_1 \\ a_2 \gg a_3 \\ a_3 \gg a_1 \\ a_3 = d_2 \\ d_1 = a_1 + a_3 \end{cases} \tag{31}$$

4. Simulation Results

Using Maple/Sim software[9], we build a two DOF mode with five different key structural parameters for the linear driving manipulator (see Table 1), after that we make a simulation for the kinematical and dynamical model.

Table 1. Five different kinds of parameters.

Model	a_1(m)	a_2(m)	a_3(m)	d_1(m)	d_2(m)	Lx^* (m)
P1	0.12	0.41	0.38	0.5	0.26	0.9021
P2	0.17	0.41	0.38	0.5	0.26	0.9525
P3	0.17	0.41	0.38	0.55	0.38	0.9600
P4	0.17	0.55	0.38	0.55	0.38	1.1000
P5	0.17	0.65	0.38	0.55	0.38	1.2

Note*: $L_x = \sqrt{(a_3 + a_1 + a_2)^2 - (a_3 - d_2)^2}$

To input the same constant yaw torque(T_1 =1000Nm) and constant pitch torque(T_2 =34000Nm) on each model, meanwhile setting the base pitch angle as 0°, 15°, 30°, 45°, 60°, 75°, 90°, also setting the base yaw angle from -90°to 0°, we plot thechanging curves of F_L and F_R with simulation from the seven different models.

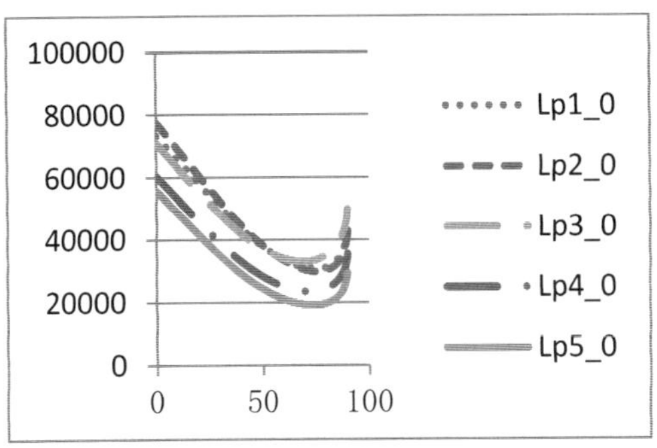

Figure 4. Changing curves of $F_L(\theta_2$=0°)

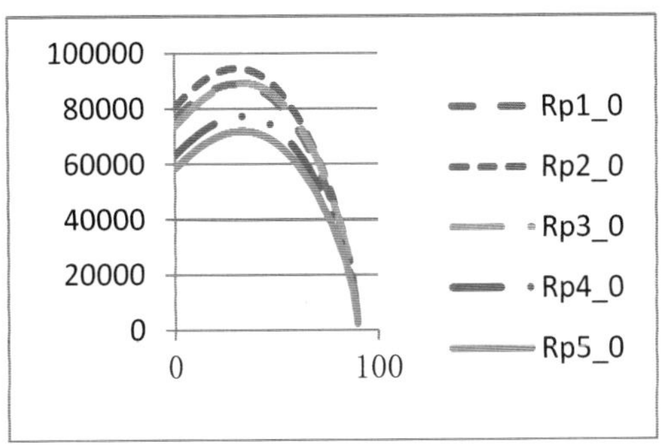

Figure 5. Changing curves of F_R (θ_2=0°).

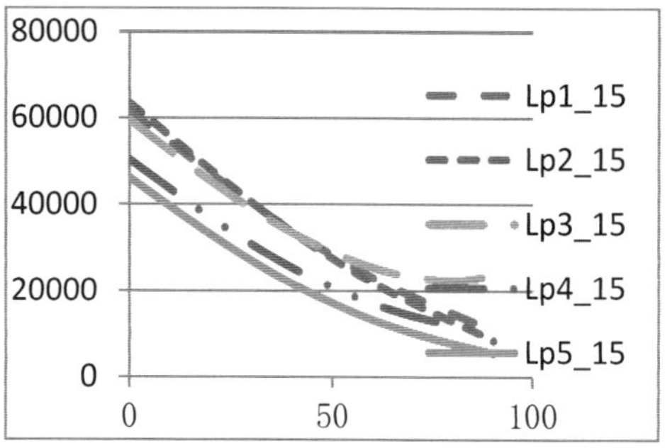

Figure 6. Changing curves of F_L (θ_2=15°).

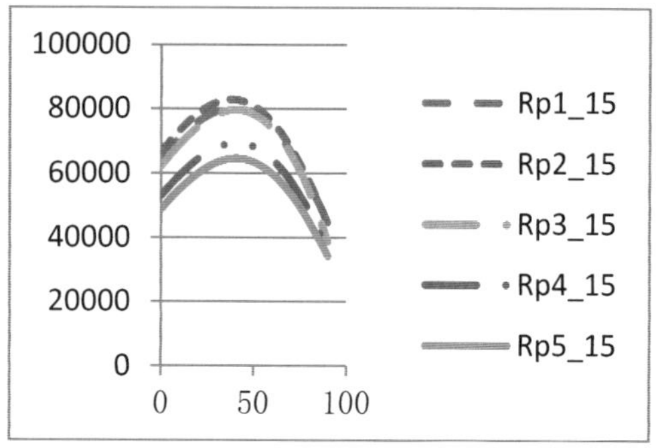

Figure 7. Changing curves of F_R (θ_2=15°).

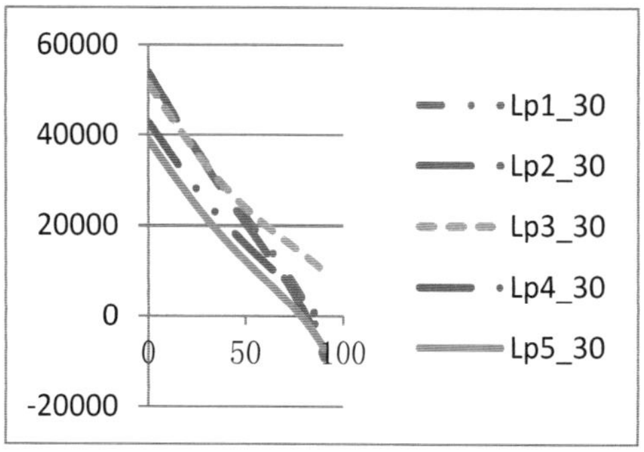

Figure 8. Changing curves of F_L (θ_2= 30°).

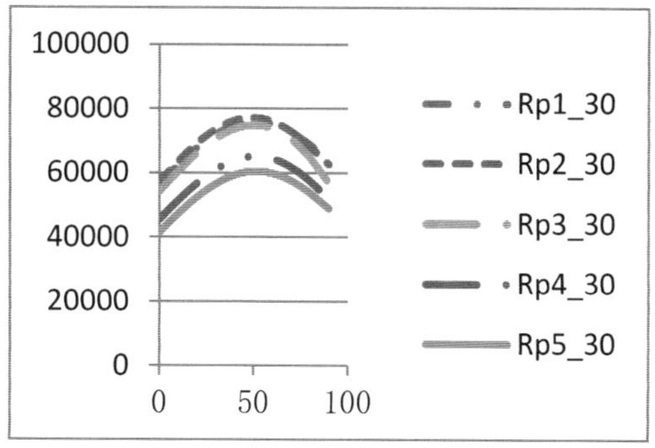

Figure 9. Changing curves of F_R (θ_2= 30°).

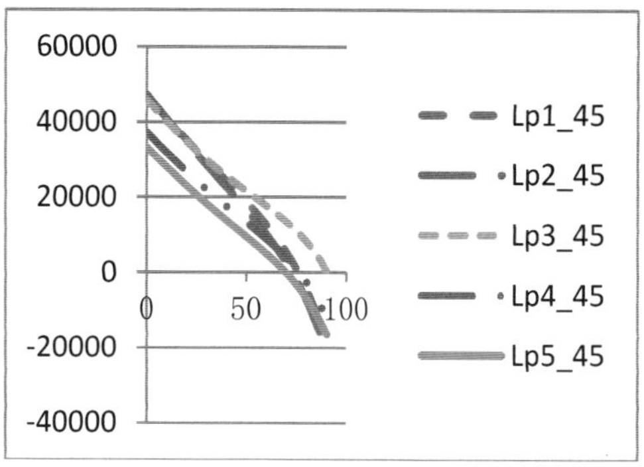

Figure 10. Changing curves of F_L (θ_2= 45°).

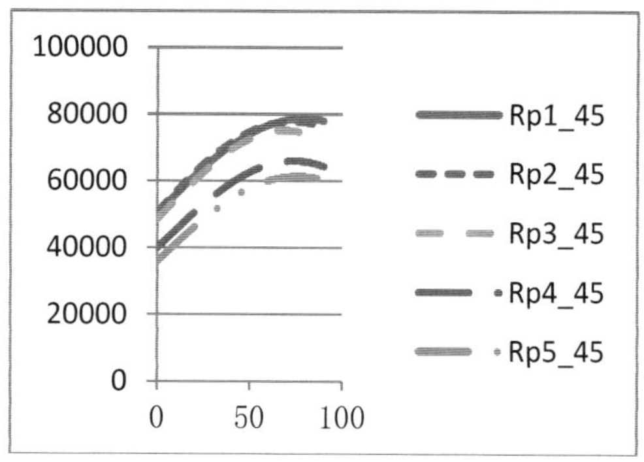

Figure 11. Changing curves of F_R (θ_2= 45°).

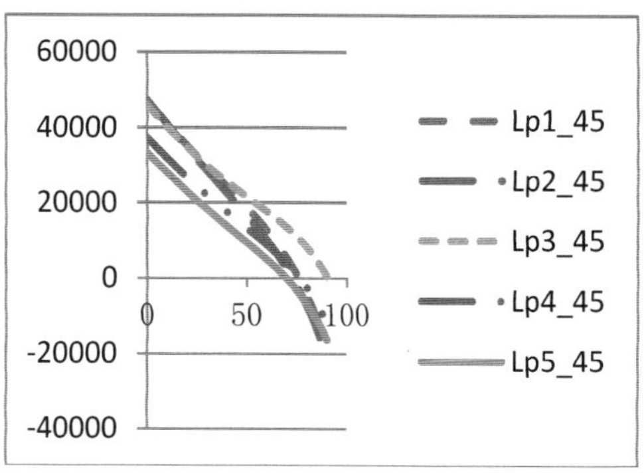

Figure 12. Changing curves of F_L (θ_2= 60°).

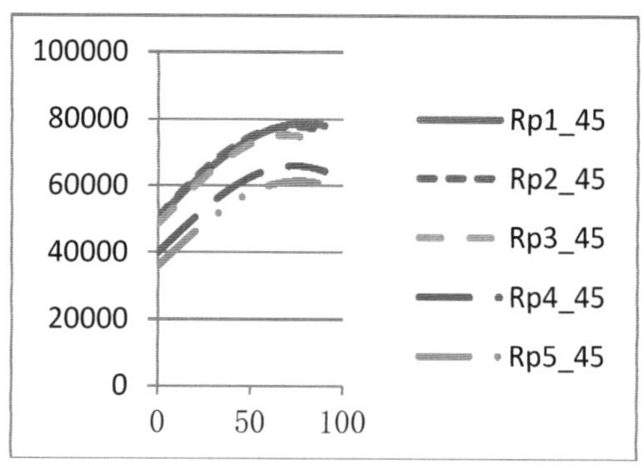

Figure 13. Changing curves of F_R (θ_2= 60°).

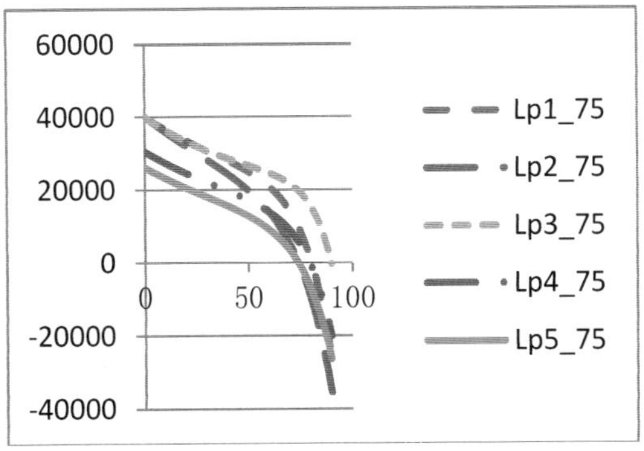

Figure 14. Changing curves of F_L (θ_2= 75°).

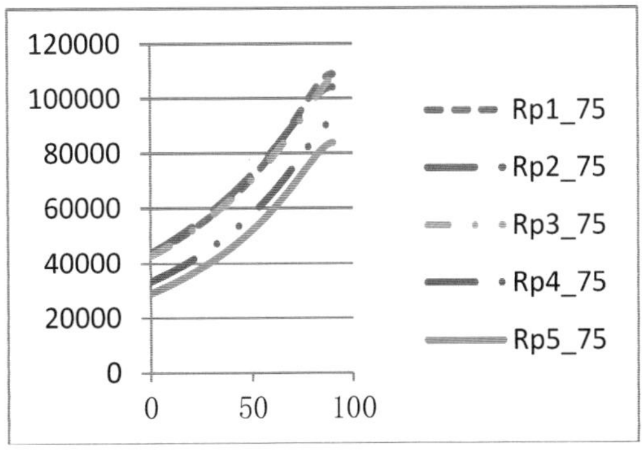

Figure 15. Changing curves of F_R (θ_2=75°).

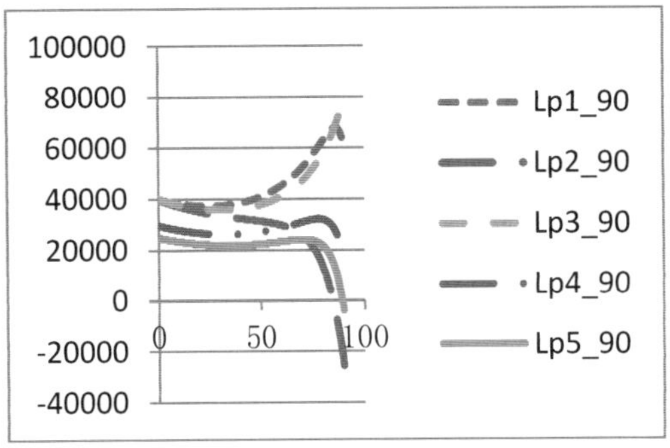

Figure 16. Changing curves of F_L (θ_2= 90°).

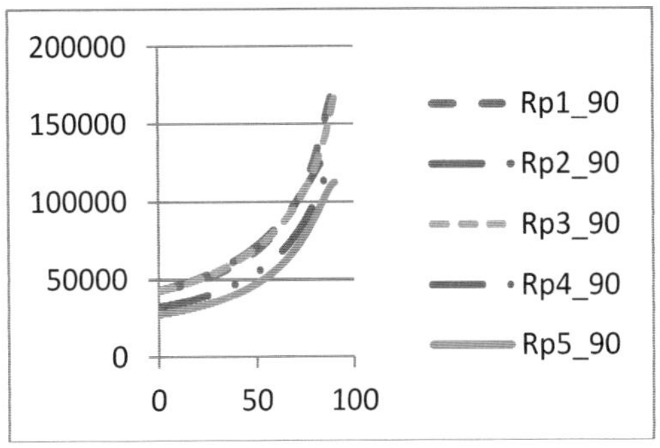

Figure 17. Changing curves of F_R (θ_2=90°).

It can be seen from Figure 4 to Figure 17,when the manipulator working in the range -90°≤θ_1≤90°, 0°≤θ_2≤90°, the bigger $a_2$is, the bigger yaw or pitch driving torque is. When the smaller $a_1$is and the closer $d_1$toa_1 +a_3 is and the closer value $d_2$to$a_3$is, thus the bigger yaw or pitch driving torque is.The simulation results are consistent with Formula (31).

5. Conclusions

In this paper, a two DOF joint mechanical design scheme for heavy-load manipulator based on linear driving is proposed, which has the advantages of high driving capacity, simple structure and light weight. Through theoretical analysis and simulation, we come to a conclusion that when setting the same loader on the base yaw and pitch joint, in order to input lower linear driving force, the parameter a2 should be as maximum as possible, a1 should be as minimum as possible, d1 should be as close to a1 +a3 as possible, d2 should be as close to a3 as possible. Compared with simulation results, the conclusions are also validated to be reasonable.

6. References

[1] Kock S, Vittor T, Matthias B and et al 2011 Robot concept for scalable flexible assembly automation: a technology study on a harmless dual-armed robot. *Int. Sym. on Assembly and Manufacturing(IEEE)* p1-5

[2] Hedelind M and Kock S 2011 Requirements on flexible robot systems for small parts assembly: a case study. *Int. Sym on Assembly and Manufacturing(IEEE)* p1-7

[3] Zhang B, Gravel D, Zhang G and et al 2011 Robotic force control assembly parameter optimization for adaptive production. *Int. Conf. on Robotics and Automation(IEEE)* p464-469

[4] Sang C, Eakins W and Trends T 2010 Opportunities for robotic automation of trim and final assembly in the automotive industry. *Conf. on Automation Science and Engineering(IEEE)* p21-24

[5] Chen H, Wang J, Zhang B and et al 2010 Towards performance analysis of wheel loading

process in automation manufacturing. *Conf. on Automation Science and Engineering(IEEE)* p234-239

[6] Hedelind M and Jackson M 2011 How to improve the use of industrial robots in lean manufacturing systems. *J. Manu. Tech. Mana.***22**(**7**):891-905

[7] Cai Z X 2016 *Fundamentals of robotics*(China Machine Press) p36-38

[8] Cai Z X, Xie B 2015 *Robotics* (Tsinghua University Press) p109-111

[9] Lu Z G, Xu J L 2012 *MapleSim System Modeling and Simulating* (China Railway Press) p70-129

IOP Conf. Series: Materials Science and Engineering **575** (2019) 012012 doi:10.1088/1757-899X/575/1/012012

Vulnerability verification for illegal attack on secure CAN communication

K Nishida[1*], Y Nozaki[1] and M Yoshikawa[1]

[1] Department of Information Engineering, Meijo University, 1-501 Shiogamaguchi, Tempaku-ku, Nagoya, Aichi 468-8502 Japan.

*150441108@ccalumni.meijo-u.ac.jp

Abstract. Message authentication using MAC has been proposed in order to prevent illegal attacks against CAN which is the standard of in-vehicle communication protocol. It also prevents replay attacks by inserting counter number in MAC. However, it is difficult to synchronize with counter number. Thus, message authentication cannot be performed correctly. Therefore, the counter synchronization mechanism is required. Regarding in-vehicle system, security verification is important as well as safety evaluation. This study proposes a new replay attack focused on the counter synchronization. This study also implements CAN communication with MAC. Experiments using implemented CAN communication show that the proposed replay attack succeeds. Experiment results indicate that the counter synchronization is a key technique to achieve secure in-vehicle system.

1. Introduction

Currently, an automobile is controlled by Electronic Control Unit (ECU). As the advancement of automobile, the number of ECUs is increasing. All ECUs are connected on Controller Area Network (CAN) which is the standard of in-vehicle protocol.

However, the security of CAN is not sufficient, and it is reported that the automobile is illegally controlled [1][2]. In 2015, it was demonstrated that the target vehicle could be controlled illegally and remotely [3]. This dreadful incident caused a massive recall. Therefore, many researches to improve the security of in-vehicle system has been reported. An intrusion detection method for detecting malicious messages is known as example of countermeasure against the attacks [4]. Besides, secure boot has been proposed to detect tampering with ECU [5]. The most typical countermeasure is to perform message authentication by Message Authentication Code (MAC). In secure CAN communication using MAC, MAC is generated including the nonce by the counter because simply generated MAC is subject to the replay attack. On the other hand, if the counter is out of synchronization, there is a possibility that authentication cannot be performed correctly. Therefore, in the authentication method proposed in [6], the global nonce generator is introduced. The nonce generator periodically broadcasts a random global nonce, and the counter is resynchronized.

Since attacks on in-vehicle system are threat to life, to enhance security of CAN is very important. Therefore, security evaluation of secure CAN communication is necessary.

This study proposes a new attack method on CAN communication with MAC. Actually, in CAN communication using counter value for MAC generation, the resynchronization mechanism of the counter is required. The proposed method performs replay attack focused on a synchronization message of the counter. Experiments using mock-up system, which is implemented CAN communication with MAC generated from control data and counter value, show that the proposed method succeeds. Experiment results also indicate that synchronization of the counter needs to be performed securely.

2. Preliminaries

2.1. Controller Area Network

CAN [7] is a communication protocol used in a vehicle network. CAN transmit data using a potential difference between two signal lines. CAN is resistant to noise because the noise applied to the two signal lines is the same and the potential difference does not change. CAN protocol adopts multi-master system, and when the CAN bus is free, communication can be started from any node. If the multiple nodes start communication at the same time, high priority message is transmitted. In the arbitration, the smaller ID has priority over the larger one. CAN communicates by using four kinds of frames called data frame, remote frame, error frame, and overload frame. Figure 1 shows the format of the data frame. The data frame is a format used in data transmission, and data is stored in the data field of the data frame. The maximum length of the data field is 64 bits.

2.2. Message Authentication Code

MAC is the information for message authentication. Message authentication is to check whether the message is tampered. MAC is generated from the shared key and the message. Figure 2 shows the overview of the message authentication. The sender and the receiver share key before starting communication. First, the sender generates MAC from the shared key and the data. The sender sends the data and generated MAC to the receiver. Then, the receiver generates MAC in the same way as the sender. If the generated MAC is equal to received MAC, the sent data is not tampered.

2.3. Cipher-based Message Authentication Code (CMAC)

CMAC [8] is MAC based on block cipher. Figure 3 shows the two cases of MAC generation process in the CMAC. Figure 3 (i) shows the case that the length of M_n fulfills the block size b, and figure 3 (ii) shows the other case. The subkeys K_1 and K_2 are generated from the shared key. The message M is divided into M_1 to M_n by the block size b. As shown in figure 3 (ii), padding is performed on M_n when the length of M_n does not fulfill the block size. MAC is generated by performing encryption ENC_k and exclusive OR operation. The generated MAC length is equal to the block size b.

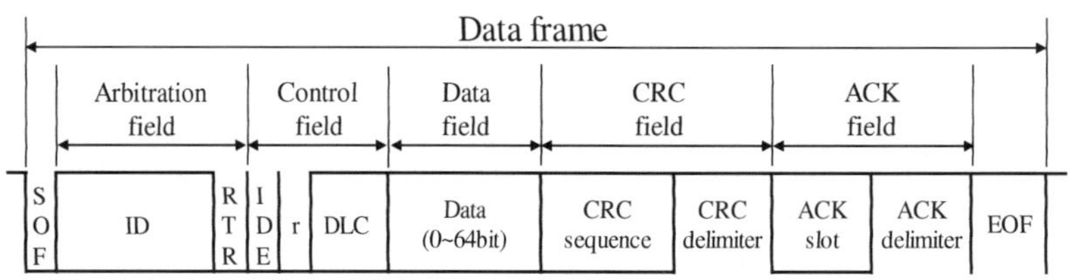

Figure 1. Data frame format.

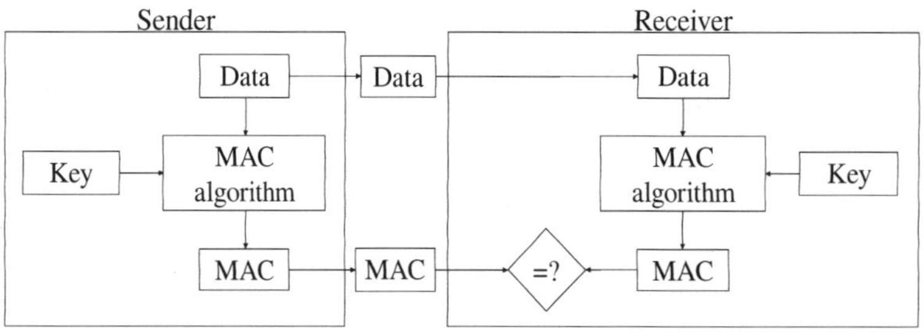

Figure 2. Overview of the message authentication.

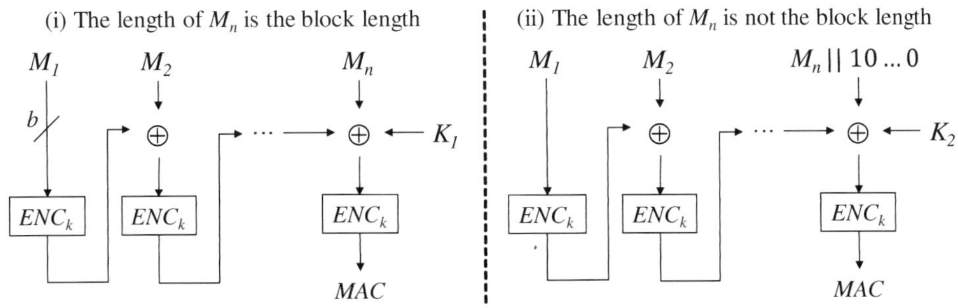

Figure 3. MAC generation processes in CMAC.

3. Proposed method

3.1. Countermeasure of in-vehicle system

This study applies AES-CMAC [9] to CAN communication. AES-CMAC is an authentication algorithm based on CMAC with Advanced Encryption Standard (AES). Figure 4 shows the overview of the implemented CAN communication with MAC. The sending side and the receiving side hold the secret key and the counter. The secret key is shared before starting communication.

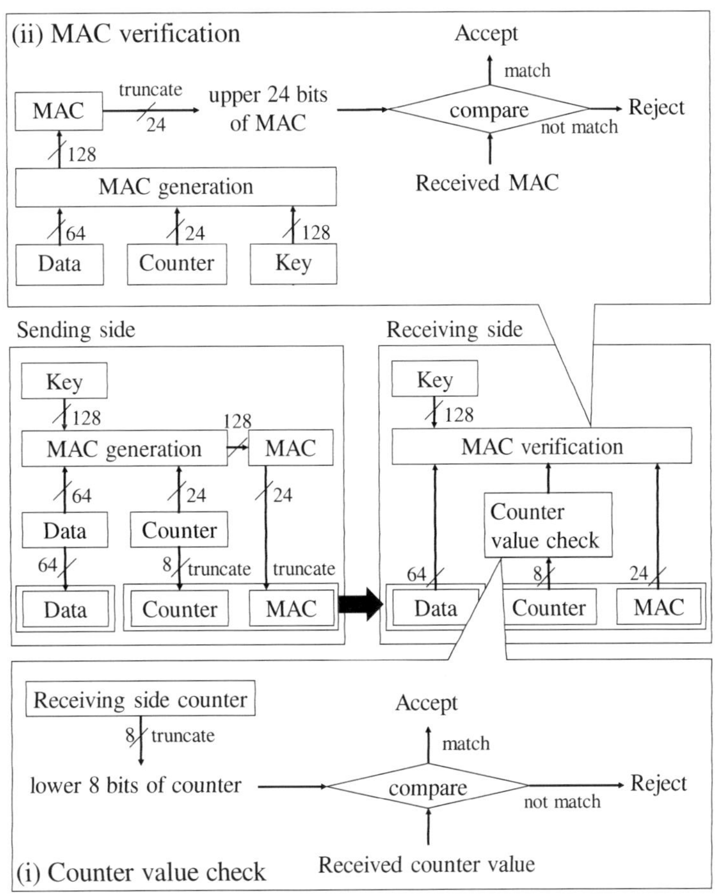

Figure 4. Overview of the implemented CAN communication.

First, the sending side or the receiving side increments each counter at the time of sending or receiving control data. Next, as shown in figure 4, on the sending side, MAC is generated from a 128-bit key, a 64-bit control data, and a 24-bit counter. At this time, the generated MAC length is longer than the data field length of data frame. Therefore, part of the generated MAC is truncated, and the upper 24 bits are treated as MAC. Figure 5 shows messages format. Then, the sending side sends the lower 8 bits of the counter value and the MAC after sending the control data. Also, the receiving side receives the 8-bit truncated counter value and the 24-bit MAC after receiving the control data. Then, counter value check part compares the received counter value and the lower 8 bits of the counter value on the receiving side. When the counter values do not match, the control data is discarded, as shown in figure 4 (i). Subsequently, the receiving side generates MAC in the same way as the sending side. Finally, MAC verification part compares the received MAC and the upper 24 bits of the generated MAC, as shown in figure 4 (ii). When the MACs do not match, the control data is discarded.
The synchronization of the counter is performed by CAN message. When the synchronization message is received, ECU sets the held counter to zero.

3.2. Attack procedure
This study assumes that an attacker can capture messages on the CAN bus and freely send messages. Figure 6 shows the overview of the proposed method. The proposed method performs replay attack focused on synchronization message of the counter. First, an attacker observes CAN bus, and captures synchronization message of the counter and subsequent messages (see figure 6 (i)). Then, the attacker sends the captured messages to CAN bus (see figure 6 (ii)). Actually, the implemented system does not perform to authenticate synchronization message. As a result, the messages retransmitted from the attacker can be authenticated because the counter is reset by the synchronization message included in captured messages.

4. Evaluation experiments

4.1. Experiments environment
Figure 7 shows experiments environment. This study used two M32C boards as ECU for sending and receiving side. The CAN communication using AES-CMAC (see Sect. 3.1.) is implemented on the M32C boards. The control data is transmitted from the controller to the sending side. In addition, the control data is displayed on the 7 segment LED on the M32C boards. In 7 segment LED, when authentication succeeds, the displayed control data is updated. This confirms that the messages retransmitted from the attacker is authenticated. For the observation of CAN messages, the MicroPecker CAN Analyzer was used.
Table 1 shows data contents of CAN messages in the implemented CAN communication. The synchronization message with ID 010 is transmitted by pushing button on the M32C board to perform easily experiments. The messages with ID 101 and 103 are control data and the messages with ID 102 and 104 are data for authentication. As the counter needs to be synchronized for authentication, the high priority ID 010 was assigned to the synchronization message.

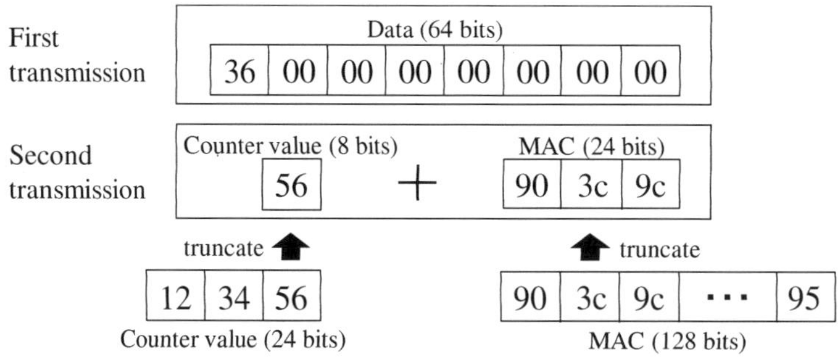

Figure 5. Message format in the implemented CAN communication.

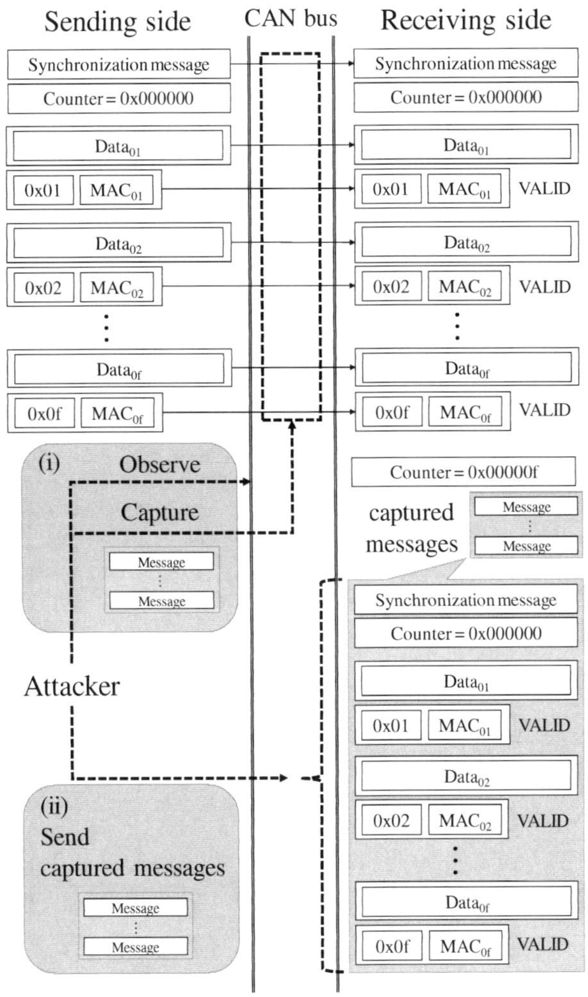

Figure 6. Proposed method.

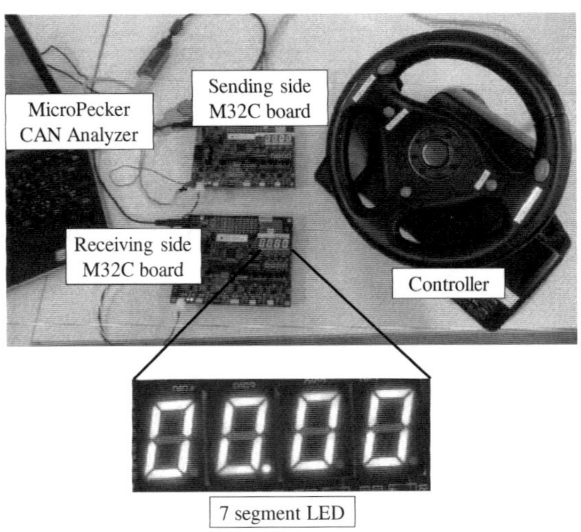

Figure 7. Experiments environment.

Table 1. Data contents of CAN messages.

ID	Length	Content
010	3 bytes	Synchronization message
101	8 bytes	Control data
102	4 bytes	Counter value and MAC with ID 101
103	8 bytes	Control data
104	4 bytes	Counter value and MAC with ID 103

4.2. Experiment results

First, this study confirmed that the implemented CAN communication communicated correctly. Figure 8 shows observation result of the implemented CAN communication. As shown in figure 8, counter value and MAC were transmitted after the control data. 7 segment LED on the receiving side displayed transmitted control data. As a result, communication was performed normally in the implemented CAN communication.

Next, the normal replay attack was conducted. In the normal replay attack, an attacker retransmitted only one set of control data, counter value and MAC. Figure 9 shows observation result of the normal replay attack. In this figure, "5C 00 00 00 00 00 00 00", "57", and "72 41 F9" represent the retransmitted control data, counter value, and MAC, respectively. On the receiving side, the retransmitted control data was not displayed on 7 segment LED because different counter values were used each time for MAC generation. Thus, the implemented CAN communication can prevent the normal replay attack.

Finally, the proposed method was conducted. Figure 10 shows the results by the proposed method. In figure 10, the message with ID 010 is the synchronization message of the counter. At this time, the counter of the receiving side was reset to zero by this message. Therefore, the retransmitted messages after synchronization message were authenticated because the counter value of the receiving side synchronized that of the retransmitted messages. Also, on the receiving side, retransmitted control data was displayed on 7 segment LED. Thus, the proposed method can succeed in illegal attack. Therefore, countermeasure against replay of synchronization message is necessary.

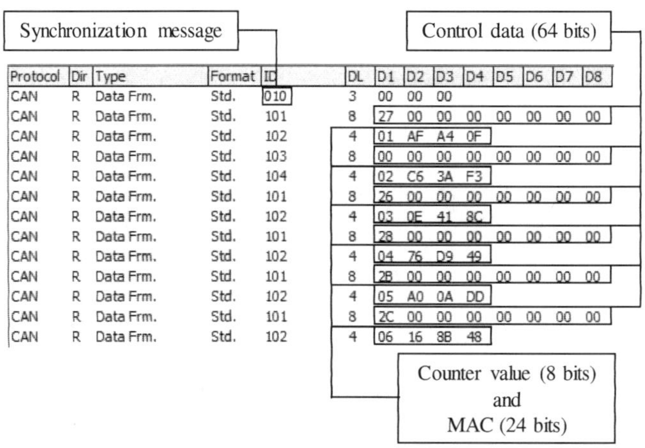

Figure 8. Observation result of the implemented CAN communication.

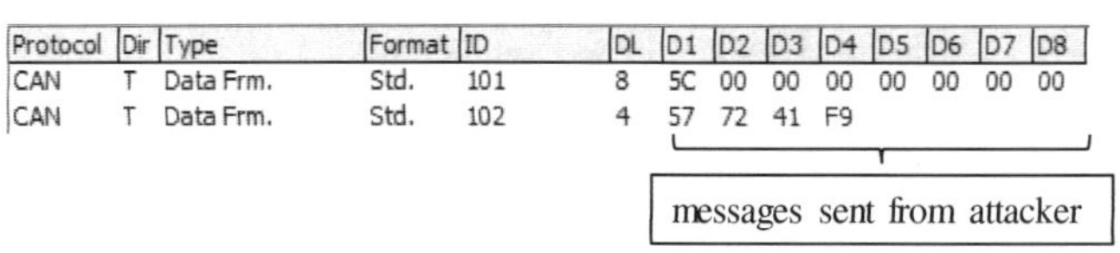

Figure 9. Observation result of the normal replay attack.

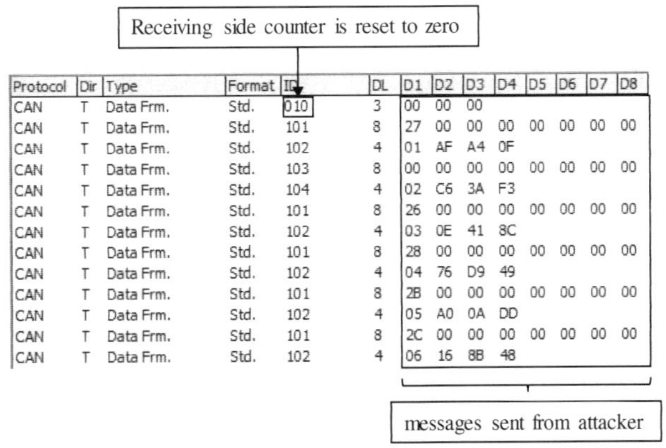

Protocol	Dir	Type	Format	ID	DL	D1	D2	D3	D4	D5	D6	D7	D8
CAN	T	Data Frm.	Std.	010	3	00	00	00					
CAN	T	Data Frm.	Std.	101	8	27	00	00	00	00	00	00	00
CAN	T	Data Frm.	Std.	102	4	01	AF	A4	0F				
CAN	T	Data Frm.	Std.	103	8	00	00	00	00	00	00	00	00
CAN	T	Data Frm.	Std.	104	4	02	C6	3A	F3				
CAN	T	Data Frm.	Std.	101	8	26	00	00	00	00	00	00	00
CAN	T	Data Frm.	Std.	102	4	03	0E	41	8C				
CAN	T	Data Frm.	Std.	101	8	28	00	00	00	00	00	00	00
CAN	T	Data Frm.	Std.	102	4	04	76	D9	49				
CAN	T	Data Frm.	Std.	101	8	2B	00	00	00	00	00	00	00
CAN	T	Data Frm.	Std.	102	4	05	A0	0A	DD				
CAN	T	Data Frm.	Std.	101	8	2C	00	00	00	00	00	00	00
CAN	T	Data Frm.	Std.	102	4	06	16	8B	48				

Figure 10. Observation result of the proposed method.

5. Conclusions

This study proposed a new replay attack focused on the counter resynchronization of CAN communication with MAC generated from control data and counter value. The proposed method can attack by control data in the retransmitted message on the condition that attacker can capture and send messages. This study also implemented CAN communication using AES-CMAC. Experiments using the implemented CAN communication showed that the proposed method succeeded and countermeasure against replay of synchronization message was necessary.

Future works include study of secure synchronization of counter.

6. References

[1] Koscher K, Czeskis A, Roesner F, Patel S, Kohno T, Checkoway S, McCoy D, Kantor B, Anderson D, Shacham H and Savage S 2010 Experimental Security Analysis of a Modern Automobile *Proc. 2010 IEEE Symp. on Security and Privacy* pp 447–462

[2] Miller C and Valasek C 2014 Adventures in Automotive Networks and Control Units https://ioactive.com/pdfs/IOActive_Adventures_in_Automotive_Networks_and_Control_Units.pdf

[3] Miller C and Valasek C 2015 Remote Exploitation of an Unaltered Passenger Vehicle http://illmatics.com/Remote%20Car%20Hacking.pdf

[4] Kuwahara T, Baba Y, Kashima H, Kishikawa T, Tsurumi J, Haga T, Ujiie Y, Sasaki T and Matsushima H 2018 Supervised and Unsupervised Intrusion Detection Based on CAN Message Frequencies for In-vehicle Network *J. of Information Processing* **26** pp 306–313

[5] Takemori K, Mizoguchi S, Kawabata H and Kubota A 2016 In-Vehicle Network Security Using Secure Element *IEICE TRANSACTIONS on Fundamentals of Electronics, Communications and Computer Sciences* **E99-A** 1 pp 208–216

[6] Nürnburger S and Rossow C 2016 – vatiCAN – Vetted, Authenticated CAN Bus *Proc. Cryptographic Hardware and Embedded Systems – CHES 2016* pp 106–124

[7] Robert Bosch GmbH 1991 CAN Specification

[8] Dworkin M 2005 Recommendation for Block Cipher Modes of Operation: the CMAC Mode for Authentication https://nvlpubs.nist.gov/nistpubs/SpecialPublications/NIST.SP.800-38b.pdf

[9] Song JH, Poovendran R, Lee J and Iwata T 2006 The AES-CMAC Algorithm https://tools.ietf.org/html/rfc4493

IOP Conf. Series: Materials Science and Engineering **575** (2019) 012013 doi:10.1088/1757-899X/575/1/012013

Improving the Absolute Accuracy by Online Interpolation Technique of Industrial Robots

Darpan Patel[1], Lars Lienenlüke[1a], Simon Storms[1], Christian Brecher[1] and Johannes Wied[2]

[1]Department Automation and Control, Chair of Machine Tools, Laboratory for Machine Tools and Production Engineering (WZL, RWTH Aachen), Campus-Boulevard 30, 52074 Aachen, Germany

[2]Technology factory, Press shop, Mercedes-Benz Plant Sindelfingen, 71059 Sindelfingen, Germany

e-mail address corresponding author: l.lienenlueke@wzl.rwth-aachen.de

Abstract. This paper describes a novel method to improve the accuracy of industrial robot by calibration technique. Robot calibration is broadly classified into model-based and modeless techniques. In this paper, a modeless technique capable of interpolating the deviation along the Z-axis is defined which doesn't consider any kinematic model, instead adapts interpolation method to compensate errors. The robot's deviation is measured with a laser tracker and the deviation controlling parameters are calculated. The controlling parameters interpolate the deviation and with an online compensation program this deviation is compensated in real time scenario. This modeless technique is experimentally tested and it demonstrates a substantial improvement in the accuracy of the robot.

1. Introduction

Precision machining is widely practiced in automobile and aerospace industries. A conventional approach for precision machining is with machine tools. They have been widely used in the industries and are preferred over the articulated robots. One of the main reasons of machine tools being chosen over robots is due to its better accuracy. While articulated robots are known for its repeatability, accuracy remains a topic of discussion. Most manufacturing industries use the robot manipulators for pick and place tasks which they perform adeptly. In comparison with the special purpose machines, the robots offer a greater work volume, flexibility and are relatively cost effective [1, 2].

Modern manufacturing industries demand a flexibility in their manufacturing operations. Since industrial robots can perform several machining operations and can easily adapt to different work volumes, they offer a great potential for flexibility in manufacturing industries. However, the use of robots for precision machining is compromised due to its inability to perform with high accuracy. Due to this, efforts are being taken to improve the accuracy of the industrial robots and to fulfil the demands of precision manufacturing. However, robots consist of a complex assembly system and a small change in kinematic model adversely affects the robot's performance. Therefore, inaccuracy of robots is intrinsic. Hence, it is necessary to understand the robot kinematics to perform calibration tasks. There are several robot calibration companies that offer calibration solutions which are often insufficient. These model based approaches mainly consider axis length, joint angle and joint position deviations, but neglect important factors like gear backlash or joint stiffness. So the considerable effort for calibration and compensation brings only small improvements in absolute accuracy. The approach that is presented in this paper leads to a significant improvement that is easy to implement. It is also difficult to validate the results of calibration offered by these companies, as there are several aspects which may be ignored during calibration. To elaborate on this issue, the calibration experiment might be conducted under a specific loading condition and therefore when different loading conditions are applied on the robot, it is difficult to comment on inaccuracy observed even after calibration. Similarly, the ambient temperature of the robot needs to be monitored and maintained to accurately calibrate a robot. Therefore, the above-mentioned aspects are to be considered to develop a compensation technique capable of being installed on several robots is to be developed.

The research work is further distributed in the following manner. The initial survey of the recent calibration technologies are explained in section 2. Section 3 presents measurement technique and the methods to estimate the amplitude of deviations automatically with Matlab and Simulink to determine the parameters required further for compensation. In section 4, the concept of the modeless calibration method developed is described. The working principle of the implemented modeless technique is explained in the section 5. The compensation technique is further verified experimentally by conducting measurements with the laser tracker and validated by performing milling process in section 6. The interpretation of the results along with the conclusion is discussed in section 7.

2. State of the Art

The robot consists of a complex assembly structure. The accuracy of the robot is usually affected due to reasons like prolonged use of robots, assembly, installation tolerances, stiffness, gear backlashes as well as software errors like calculation error of the TCP (Tool Center Point) position inserted in robots control system. Finding the location of such errors is a great challenge. Therefore, altering the design or disassemble and reassembling the robot can create new challenges. Different techniques and methods have been developed to calibrate robots which is based on different models. In an effort to further elaborate calibration techniques performed on robots, it is broadly classified as model-based and modeless methods [3]. The model-based calibration technique encompasses implementation of a kinematic model for the robot, conducting measurements by moving the robot inside the work volume and then with the modified kinematic parameters, compensation is performed [4]. The advantage of a model-based calibration is that the kinematic parameters can be determined for a relatively large work volume. All poses lying within this working volume can be compensated with these parameters [3, 5]. This implies that as the work volume is reconfigured, a new set of parameters need to be determined. The disadvantage of this model is that one requires a thorough knowledge of kinematical modeling as well as determining its parameters is a complex process. This knowledge might remain limited to few experts in kinematical modeling and is not easily interpretable for field engineers [3, 5]. On the contrary, a modeless method does not consider any kinematic modeling and parameter estimation. In this method, measurements are taken throughout the robot's workspace and the errors are recorded. With the help of these errors, the robot's workspace is divided into a number of 2D squares and 3D cubes. The errors along these 2D squares and 3D cubes are known from previous measurements and errors can be compensated by simple interpolation techniques [6]. Even though this method consists of a relatively large measurement, its simple implementation and effectiveness have led many industries to adopt this technique [3, 5].

Since the use of robots in machining industries has increased, several researchers have taken up the challenge to develop accurate kinematic models. A linear model and complex model were developed which could accurately compensate the first-order and second-order error effects. Since then several kinematic models such as Hayati et al. models, Stone and Sanderson's S-model and Zhuang et al. model have been developed [4, 7]. But the most recent research has led to the implementation of the Screw method and the Denavit-Hartenberg (DH) method out of which DH method is widely practiced in the industries. Based on the short comings of DH method and the complexity involved in matrix calculation, an alternative method called Direct Coordinate System method is developed [8]. A common approach to all the above-mentioned methods is to accurately develop a kinematic model based on the robot design and determining the optimized parameters based on the errors. With the help of these optimized parameters, a backward transformation is carried out to redefine the robot joint angles.

A common approach for measuring the accuracy of the robot is with a laser tracker or by a stereo camera. Normally, a single reflector is attached to the spindle to record its coordinates. However, this imposes a few restrictions to move the TCP in certain positions and poses. This can be overcome with a square tool consisting of 4 reflectors on each side. Hence ensuring that at least one reflector is detected by the laser tracker. Subsequently, the TCP can be guided to several positions and with different poses. By conducting several calibration tests with this tool, the kinematic parameters of joints 2, 3, 4 and 5 can be determined along with the present compliance [9]. Another method to evaluate the DH parameters is by the stereo vision system. By placing 2 CCD cameras perpendicular to one another, a virtual TCP can be generated. This virtual TCP is moved along the measurement points manually within the range of calibration volume. A machine vision algorithm is developed to determine the robot joints and to further rectify these joints [10]. Another approach to finding the optimized joint parameters is to place a marker on the TCP which can be detected by 2 or more cameras placed at different positions [4].

On the other hand, the modeless approach does not consider any kinematic model, instead, it adapts interpolation methods to compensate error. Several techniques developed to compensate error with modeless approach were found in the literature. The TCP is moved along the work volume and the position of it is continuously measured with a CMM (Coordinate Measuring machine) device. The work volume of the robot is divided into a number of 2D and 3D cells and the error measured along these cells is noted [6]. A method is developed to statistically analyze the error and to develop a histogram function based on the data. An algorithm is developed with this histogram function to improve the performance of robot [11]. Another method is developed by dividing the workspace into fan shaped cells with each cell consisting of 8 points. In this technique, the error along each point of the cell is measured and the position of the target point is estimated by linearly interpolating the position based on the error calculated at the grid points [6]. Subsequently, the interpolation of the target points is proposed and implemented with fuzzy logic. A fuzzy logic incorporates a membership function for each of the measured output and desired target output. The difference between the two-membership function is defined by the error calculated. A set of control rules are required to be defined to determine the position error. That implies, to interpolate the error in a cube defined by 8 points, 8 control rules are required [5].

3. Experimental Setup

In this research, the authors have developed a modeless calibration technique which incorporates a work volume of 1000 mm×4200 mm×500 mm. The deviation along Z-axis within this work volume is measured with a Leica laser tracker. The laser tracker is a highly accurate measuring device with an accuracy of ±15 μm (+ 6μm for high distances). The robot used for the experimentation is a KUKA KR500 L480-3 MT. These measurements are further processed in Matlab to determine the inaccuracy of the robot. Further, with the help of Simulink models, parameters controlling the inaccuracy are determined. With these parameters, the error is interpolated based on the TCP position and further compensated. The causes of deviation and its effect on machining primarily focused on the issue of deviation observed in the Z-axis. The deviation along Z-axis accounts for deteriorating the surface quality during machining operations like milling and laser processing. It is observed that the error present in first 3 joints contributes a higher percentage of error in comparison to the errors of the other 3 joints [6]. Based on the motion of the first 3 joints of a robot, a technique to determine the amplitude of this deviation is developed.

3.1. Measurement technique

A curved path test is defined which takes into account the work volume of the robot during machining (refer Figure 1). The test is necessarily performed by guiding the TCP along a path defined by 24 points in one plane. During the measurement the laser tracker is placed stationary at a corner of the robot cell. To completely encompass the whole work volume, joint 1 is rotated from +50° to –50° by a constant increment of 25°± 5°. During this motion of joint 1, the reflector needs to be continuously rotated for the laser tracker to detect it. The first curve is measured by rotating the joint 1 with a radius of X = 2400 mm. In all, 5 curves are measured with each curve exceeding the previous one by 200 mm in positive X-axis. Whereas, the limits of Y-axis are approximately ± 2000 mm. The measurement is carried out for each plane's starting from Z = 500 mm to 800 mm. The defined measurement volume is preferred since the machining operations are conducted within the illustrated volume. As previously mentioned, the errors of joints 1, 2 and 3 are to be measured. For this reason, the joints 4 and 6 are not changed at all and the spindle is maintained perpendicular to the plane of measurement. Further, the joint 5 moves itself by a minimal angle as compared to the first 3 joints to maintain the spindle perpendicular to the plane of measurement.

Further, a 6-points slant test is illustrated by 6 points positioned diagonally to one another. This test is conducted only in the positive Y-axis once symmetric deviation is observed along negative and positive Y-axis. Additionally, all the 6 points cannot be measured in the negative Y-axis since the laser tracker is placed in this corner of work volume. Hence, points 5 and 6 would not be detected in negative Y-axis by the laser tracker as they lie too close to the measuring volume of the laser tracker. The first point is measured at X = 2200 mm and Y = 0 mm. Subsequent points are measured by gradually travelling in Y-axis and increasing distance in X-axis by 200 mm in comparison to the previously measured point. The limit of Y-axis is gradually increased from 0 to 2000 mm. The measurement is carried out for Z = 500 to 800 mm.

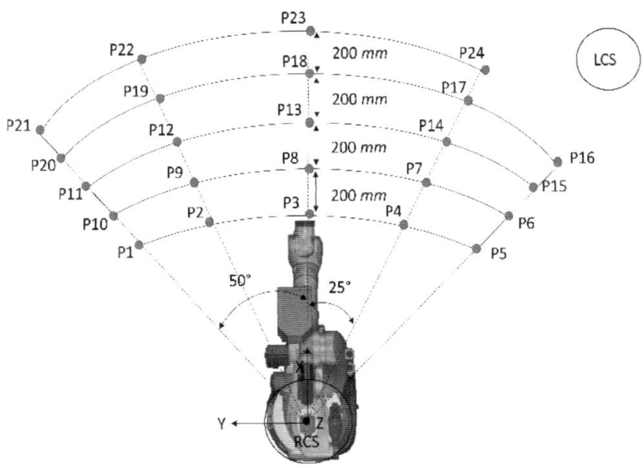

Figure 1. Curve path test

3.2. Data evaluation

The points measured by the laser tracker are measured with respect to the laser coordinate system (LCS). Therefore, to accurately evaluate the measured data, the points measured should be represented in the robot coordinate system (RCS). A Matlab program is written which helps perform the coordinate transformation. During the test, the TCP travels to different points in the defined work volume. The coordinate transformation of these points is calculated from LCS to RCS. A Matlab function performs a complex calculation to determine translation, rotation, scaling and reflection matrix with the best fit method and stores this data for further calculation. In addition, a Matlab function determines the quality of transformation by displaying the standard deviation between the theoretical and measured point. Subsequently, the measured coordinates in RCS are used for further calculation to identify the parameters affecting the deviation. To elaborate on this issue, different Simulink models are created based on the type of deviations. These Simulink models calculate the controlling parameters which are required for the compensation technique.

3.3. Types of deviation

A curved path test is initially conducted to measure the deviation along every plane. After the evaluation of the measured data in Matlab it is evident that the TCP does not travel along a straight plane. Subsequently, after an initial survey of evaluated data, it was observed that the deviation in the positive Y-axis is greater than the deviation in negative Y-axis. This phenomenon was common to all planes. The difference of deviation in Z-axis is observed by comparing the deviation of a point in positive Y-axis with its corresponding point in negative Y-axis. Hence, this confirms the presence of inclination error.

In addition to inclination error observed, a considerable deviation is observed when the TCP travels along the X-axis. When the TCP moves further from one curve to another, a radial distance of 200 mm is added to the previous radius and ensures that each curve is equally spaced to one another. After comparing the deviations observed for all points in one curve with their corresponding points to the adjacent curve, it is observed that the deviation gradually increases for each curve. This deviation is referred as the forward deviation. Furthermore, as the TCP travels in the positive and negative Y-axis along the curve, the deviation increases forming a curved surface which is denoted as sideway deviation.

In addition to the forward and sideway deviation, a non-linear pattern is observed while comparing the deviation for each plane. This implies, that the deviation is found to be above the intended plane of travel and at times below the intended plane of travel. A general observation for the deviation of all planes demonstrates that the deviation is negative to the expected plane till Z = 500 mm and below whereas, the deviation is positive to the expected plane from Z = 700 mm and higher. In the plane between 500 to 700 mm, a mixed deviation of positive and negative to the expected plane is observed. Furthermore, not only the deviation is either positive or negative but also the amplitude of deviation varies for every plane.

It should be noted that all the tests were conducted under same loading conditions and therefore different loads cannot be attributed to the cause of such a deviation. However, it can be observed that as the robot arm extends and contracts to travel along different curves, the force acting at the end effector might enhance the effects of joint clearance and drive backlash. This phenomenon of deviation can be attributed to joint clearance and drive backlash of the joint 2 and 3 which are altered for each curve. Therefore, the curve path test is capable of evaluating 3 different types of errors along X-axis (forward deviation), Y-axis (sideway deviation) and Z- axis (positive and negative deviation). The amplitude along X and Z-axis can be appropriately determined by this method. However, to determine the amplitude of sideway deviation the 6-points slant test is conducted.

4. Error Compensation Based on Modeless Method

A modeless compensation, as mentioned in section 2 divides the work volume of the robot into several parts and the error for each part is noted. A similar method is implemented by conducting measurement tests in accordance with section 3.1. The curve path test conducted in a single plane measures the error in a 2D plane along X and Y-axis. The test is further conducted by changing only the coordinates of Z-axis and the coordinates of the X and Y-axis are unchanged. This ensures the error for a point with the same X and Y coordinates is obtained at different planes. In accordance to that, an error along a confined 3-dimensional volume is measured. This measurement technique enables to examine the behavior of the robotic arm and its corresponding deviation along different work volumes. Based on the deviations measured, a prominent modeless compensation approach can be adopted. A traditional approach to implement a modeless compensation is by implementing various interpolation techniques.

4.1. Parameter estimation

A huge deflection is observed known as inclination error between the points measured to the left of the robot and to the right. This deflection is observed throughout the work volume. Further, the amplitude of the deviation also remains almost constant for all the measured planes. Therefore, the slope of inclination can be calculated between the points to compensate the error. To accurately calculate the amplitude and the slope of inclination, a Simulink model is constructed. The Simulink model determines this difference along all the plane and computes an average of the difference. The output of the Simulink model is further used in the Kuka Robot Language (KRL) program to interpolate the difference. Since this inclination error is observed along the Y-axis the angle is calculated based on the distance travelled along Y-axis in the work volume.

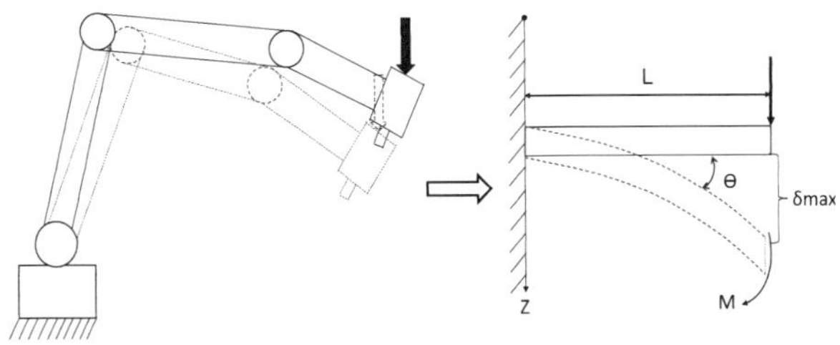

Figure 2. Graphical representation of robotic arm as cantilever beam

A common phenomenon observed along all the planes is the forward deflection along X-axis and the sideway deflection along Y-axis. This phenomenon can be compared to deflection observed in a cantilever beam (refer Figure 2). In the presence of different geometric positions and under variable loading condition as well as gravity effect, the robotic manipulator experiences a deformation [12]. In this case, the deformation is attributed to the deflection caused in the Z-axis. Subsequently, as the robotic arm extends gradually along its work volume, the weight of the end effector under the effect of gravity causes an additional deflection [13]. The results of a curve path tests reveal that under a constant load due to self-weight of the robot end effector, a forward and a sideway deflection is obtained for each curve. To accurately determine this deviation as the robot arm extends, the deviation is calculated by an equation of cantilever beam acted upon by the moment. All the robot joints are transformed into a free body diagram representing a cantilever beam. However, the measured deviation is a known factor in this case, as well as the length of the beam can be compared to the stretch of the robotic arm. Therefore, the only decisive factor to be resolved is the θ which is known as the angle of deflection. To determine θ by an equation of cantilever beam acted upon either by a force or moment is inconsequential. But this equation should be later contemplated to interpolate the deviation with KRL programming. In accordance to that, the θ is determined with a moment equation implemented with a Simulink model. The Simulink model computes a value for theta for the individual plane and further calculates an average value of theta for all planes. With this method, two different θ's are computed for each sideway and forward deflection. The curve path test is prescribed to determine the forward theta. Whereas, the 6-points slant test is performed to calculate the sideway theta.

As the robot travels further along the Z-axis the motion of robot joint 2 and 3 causes a non-linear deviation. To summarize the behavior of error, the error for lower planes is negative and the error escalates perpetually in positive Z direction for higher planes. This implies that the TCP travels either above or below the designated plane of travel. Furthermore, the amplitude of deviation for each plane also varies. This phenomenon is observed even when the measurements are conducted after compensating the inclination error, forward and sideway deflection with the KRL program. To compensate such an error, a method is established with which the robot TCP travels along the designated plane with minimum deviation along the Z-axis. A Simulink model is designed to determine the amplitude of error for each plane and to determine a compensating value for the error. For this, the curved path test is performed after compensating the inclination, forward and sideway deflection. Only the coordinates of Z-axis for all the measured planes are fed into the Simulink model as the input. The average deviation of each plane is calculated in the model. Further, the average of one plane is divided with an average deviation of the subsequent plane. The output of this calculation ensures a multiplying factor is computed between two points which acts as a slope of inclination between the errors of two planes. This multiplying factor is treated as an angle to define the slope of deviation between the two points. The number of planes to be considered for calculating the multiplying factor reckons on the behaviour of the error observed. It should be noted that if the amplitude of error for two planes is approximately same, the value of multiplying factor would eventually be reduced. This is because the slope of inclination between the two points would be less. At times, when the multiplying factor has a relatively low value, the error is not accurately compensated for the planes displaying a considerable error.

4.2. Compensation of error by interpolation

With the angle of inclination calculated for inclination error, the amplitude of deviation can be interpolated in real time with a KRL program. This deviation can be obtained by:

$$\Delta = Y_{present} \times \tan\left(\theta_{inc}\right) \tag{1}$$

where, $Y_{present}$ is the Y coordinate of the point to be interpolated and θ_{inc} is the inclination slope. This interpolated deviation given by Δ is further added to the corresponding Z coordinate of the point (refer Figure 3 (1)). Furthermore, the amplitude of deviation depends upon the coordinate of Y-axis and this deviation is added to the present coordinate value of the Z-axis in real time scenario. Since the deviation of errors was found to be greater along positive Y-axis as compared to the negative Y-axis, the inclination error compensation is implemented along the positive Y-axis to maintain a proportional deviation along positive and negative Y-axis.

As mentioned in section 4.1, two different thetas are computed for each forward and sideway deflection. The interpolation of the forward deviation in KRL is represented as:

$$L = sqrt\left(\left(X_{present}\right)^2 + \left(Y_{present}\right)^2\right) \tag{2}$$

$$\Delta_f = \frac{\left(L \times \theta_{forward}\right)}{2} \tag{3}$$

$$\Delta_s = \frac{Y_{present} \times \theta_{sideway}}{2} \tag{4}$$

where, L is the length of the cantilever beam corresponding to the robotic arm and $X_{present}$ is the X coordinate of the point to be interpolated. Further, $\theta_{forward}$ and $\theta_{sideway}$ are forward and sideway deflection slopes computed previously. Once Δ_f and Δ_s are computed, the interpolated deviations are added to the corresponding Z coordinate of the point (refer Figure 3 (2)).

Once the Simulink model computes the value of inclination slope for non-linear deviation, the cosine of the slope value is multiplied with the Z coordinate of the designated plane of travel. Herewith, an amplitude of error for the following plane is interpolated which can be either added or subtracted to the actual Z coordinate depending on the form of error measured previously (refer Figure 3 (3)). The interpolation is computed as follows:

$$\Delta = Z_{present} - \left(Z_{present} \times \cos\left(\theta_{is}\right)\right) \tag{5}$$

here θ_{is} is the inclination slope calculated to compensate the non-linear deviation. It should be noted that all the above-mentioned interpolations are calculated for each point defining the motion of the robot.

5. Compensation Working Principle

This research work is focused towards developing a compensation technique without utilizing any external control device. Based on these limitations, a method to compensate the deviations in real time scenario is developed. A common practice in the industry to machine with an industrial robot is by programming the operation in CAM (Computer Aided Machining) software like Mastercam and developing a code from a post-processor software like Robotmaster. Further, the post-processor creates program referred as main program which can be directly run in the robot controller to perform machining operation. A compensation program is developed in KRL by integrating the available syntaxes for a KR C4 controller. Further, this compensation program is presumed to perform an interpolation utilizing the parameters estimated with the Simulink. Consequently, a function is defined which is capable of performing the necessary calculation and returning the calculated value in a specified format defined in the function. It is noted that the input and output format for a function should be defined in the same format. Subsequently, a function with name CorrectPos is defined with its input as PresentPos in E6POS format. To make PresentPos be accessible for subsequent calculation, it is declared as an object or variable in E6POS. The E6POS is a data type to represent a point in cartesian workspace along with the orientation angle of the robot. This implies in addition to X, Y, Z coordinates the corresponding A, B and C angles are defined. Further, this data type consists of S (Status) and T (Turn) which denote the axis precisely [14]. The input variables for the function are transferred from the main program. Every program generated by Robotmaster is treated as the main program. This main program consists of every point prescribed by either LIN, PTP or CIRC syntaxes. Motion programming is defined with these syntaxes. These syntaxes include the coordinate information associated with the defined points. The main program generated by Robotmaster is usually represented in LIN format as:

LIN {X 100, Y -20, Z 250, A -180, B -0.1, C -180}

In order to transfer the parameters from the main program to the developed function, the abovementioned point should be represented as:

LIN CorrectPos ({X 100, Y -20, Z 250, A -180, B -0.1, C -180})

This alteration enables the coordinate data to be transferred into the function for further processing. Once the coordinates are transferred into the function, they can be accessed by PresentPos variable. Further, the interpolation of error is executed in the following order:

1. Inclination interpolation
2. Forward and sideway deflection interpolation
3. Non-linear deviation interpolation

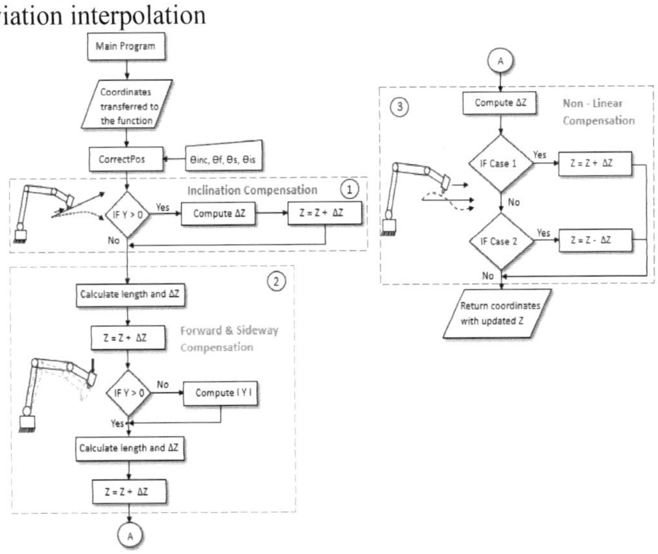

Figure 3. Flowchart for KRL compensation program

The *CorrectPos* function calculates the necessary deviation and updates the Z coordinate for every single point defined in the main program. The Z coordinate is updated in the PresentPos and the function concludes by returning PresentPos in E6POS format to the robot controller. The compensation function is so defined that the KR C4 controller first calculates the compensation and then travels to the point with the updated coordinates.

6. Validation

The compensation function developed had to be tested to check the functionality. This was performed with the aforementioned curve path test as per section 3.1. In addition, this test was performed to compare the results with and without the influence of compensation. Therefore, the measurement was carried at the same planes without changing any of the coordinates. The advantage of this procedure is, due to good repeatability properties of the robot, the deviation at each point can be distinctively observed. The amplitude of deviation after every compensation was noted, and it was found that the error compensated at one stage has an adverse effect on the parameters controlling the successive compensation. This implies, in case the inclination error is over compensated the successive sideway compensation parameter is affected. The sideway theta would reduce in this case. Similar observations were noticed between the forward and non-linear deflection.

Table 1. Deviation compensated in mm after each interpolation for planes along Z-axis

Plane	500 mm	600 mm	700 mm	800 mm
Without	-1.604	-0.992	0.973	1.52
Inclination	-1.488	-0.828	1.064	1.565
Forward & Sideway	-1.176	-0.516	0.672	1.87
Non-linear	0.076	0.46	0.144	0.584

After comparing the measurement results with and without the compensation program, it is evident the compensation function substantially compensated different types of deviations observed previously (refer Figure 4, Figure 5).

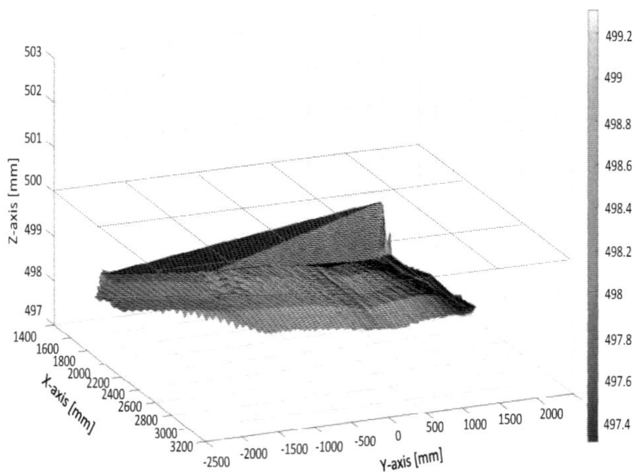

Figure 4. Measurement result without compensation for plane Z=500 mm

The accuracy of the compensation function is to be tested while performing machining operations. For this reason, a milling operation is chosen to validate the performance of the KRL program. A Zinc aluminum alloy workpiece was used on which the milling operation was performed. In order to validate the results accurately, it was necessary to perform milling operations with robot arm as extended as possible. This would ensure that the deviations in Z-axis would be greater when the robot arm was further extended. Therefore, a workpiece with dimensions 1550 × 2795 × 510 mm was selected. Further, locking pins with 50 mm height were placed between the working table and the workpiece and this changed the Z dimension to 560 mm. The workpiece was placed as far as possible in the X-axis.

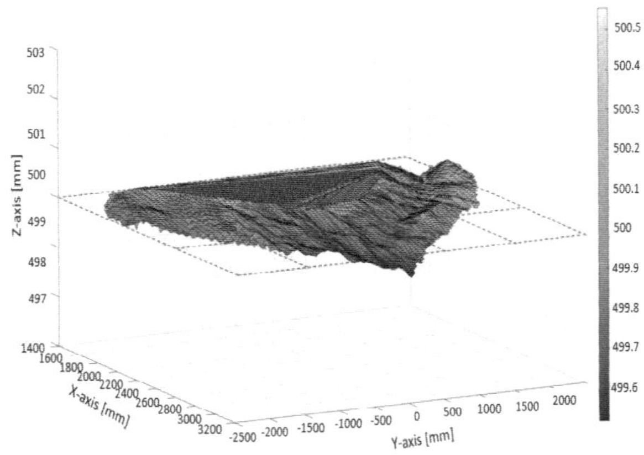

Figure 5. Measurement result with compensation program for plane Z= 500 mm.

Two paths were defined to mill the workpiece along a straight line starting from the extreme left of the workpiece and travelling towards the right with a depth of cut 0.5 mm. Both the paths were defined as far as possible in the X-axis so that the robot arm is extended. One of the paths was programmed with the compensation whereas, the other path was programmed without compensation. Further, two U-section paths are defined, each to the extreme left and right of the workpiece with respect to the robot. The path defined to the left of the robot was programmed without the compensation whereas, the path to the right was defined with the compensation. The idea was to observe a deviation between the depths of cut for both the paths. The milling operation was programmed to perform a depth of cut of 1 mm for both the paths. The milling process was conducted with the same tool and constant feed-rate for all the paths. Later, the depth of cut was measured with an absolute height gage to compare the effect of a milling process with and without compensation. Nevertheless, a substantial difference between the depths of cuts is observed between the path 1(compensation) and 2(without compensation) as well as path 3 (without compensation) and 4 (compensation). The depth of cut was measured along different points of the path to observe a pattern of deviation (refer Figure 6).

Figure 6. Validation setup

The measurements observed for path 2 confirmed the anticipated behaviour of the error when traveling from left to right. The depth of cut was considerably deeper on both the sides of the workpiece as compared to that at the centre. After comparing the deviation at the corners with the deviation at the centre it can be confirmed that the spindle travelled along a curved path as this phenomenon was demonstrated by the laser tracker measurements. In contrary to that, the depth of cut was observed to be uniform along the path 1. Theoretically, due to the extension of the robot arm to travel to path 1, the depth of cut was expected to be deeper. As for the curve path test, every curve was shifted by 200 mm and a deviation for every curve was observed in this case. However, the compensation function fairly maintained a uniform deviation along the path 1. Points corresponding to the path 2 were measured along path 1 to compare the results and observe the effects of compensation.

Comparing the results of path 1 and 2, it is evident that the depth of cut for path 2 is greater and non-uniform as compared to the depth of cut of path 1. However, the programmed depth of cut was 0.5 mm and the measured depth of cut for both the paths confirmed the presence of deviation.

Table 2. Comparison of depth of cut for path 1 and 2

Path	Right corner	Center	Left corner
1	-1.59 mm	-0.88 mm	-1.46 mm
2	-0.67 mm	-0.86 mm	-0.41 mm

To further investigate the deviations in the area near to the robot, milling across path 3 and 4 were conducted. The parallel paths defined for the U section were maintained at a distance of 350 mm apart. This distance has no significance in terms of the functioning of the compensation program but, the reason was to observe a difference in depths between the two parallel paths. Nevertheless, paths 3 and 4 were programmed symmetrically about the centre of the workpiece to compare the results. Once the milling process was performed, the depth of cut along the path was measured. For path 3 as anticipated, a difference between the depths of the cut was observed for the two parallel paths in the U-section. This implies, that as the robot arm extended gradually while milling across the U-section, the deviation in Z-axis could be clearly observed in terms of depth of cut. The measurements were conducted on the path 4 to compare the results. It was observed that the depth of cut was nearly uniform for the parallel lines. This was tested by measuring several points along the path and not only the corresponding points measured about path 3. Therefore, this uniform depth of cut can be attributed to the functioning of the compensation program.

7. Conclusion

The aim of this research work is to investigate the possibilities of developing a technique to compensate the deviation in Z-axis using industrial robots in machining operations without the application of any external controller. For this, the deviations were measured with a laser tracker and depending upon the pattern deviations a compensation program is created in KRL. Later, measurements were conducted with the compensation, and results were compared to the initial measurement. It was evident that the absolute accuracy was substantially improved from ± 1.19 mm to ± 0.47 mm in terms of standard deviation. Validations were performed experimentally to examine the functionality of compensation technique. For this, milling process was performed on a workpiece with and without the compensation function. The depth of cut for all the milling processes was compared and the deviation in depth of cuts with and without compensation was compared. The results of laser tracker measurement and validation displayed a substantial improvement in the absolute accuracy by almost 60 %. However, the reaction forces during cutting operations or the vibrations were not taken into account during validation and future work should include a method to compensate these forces. Further, the errors caused due to an external linear axis are also not considered during this research work and should be considered for future work.

8. References

[1] Lienenlüke L, Gründel L, Storms S, Brecher C. Model-based process planning for milling operations using industrial robots. In: International Conference on Control and Robotics Engineering; 2018, p. 37–44.

[2] Ceccarelli M, Glazunov VA. Advances on Theory and Practice of Robots and Manipulators: Proceedings of Romansy 2014 XX CISM-IFToMM Symposium on Theory and Practice of Robots and Manipulators. 2014th ed. Cham: Springer International Publishing; 2014.

[3] Bai Y, Zhuang H. Modeless robots calibration in 3d workspace with an on-line fuzzy interpolation technique. In: Wieringa P, editor. IEEE International Conference on Systems, Man & Cybernetics theme: Impacts of emerging cybernetics and human-machine systems conference proceedings. Piscataway (N.J.): IEEE; op. 2004, p. 5233–5239.

[4] Zhang X, Song Y, Yang Y, Pan H. Stereo vision based autonomous robot calibration. Robotics and Autonomous Systems 2017.

[5] Bai Y, Wang D. On the comparison of an interval Type-2 Fuzzy interpolation system and other interpolation methods used in industrial modeless robotic calibrations. In: IEEE International Conference on Computational Intelligence and Virtual Environments for Measurement Systems and Applications - CIVEMSA: July 27-29, 2016, Hotel Novotel Budapest Centrum, Budapest, Hungary proceedings. Piscataway, NJ, Piscataway, NJ: IEEE; 2016, p. 1–6.

[6] Guo Y, Yin S, Ren Y, Zhu J, Yang S, Ye S. A multilevel calibration technique for an industrial robot with parallelogram mechanism. Precision Engineering 2015.

[7] Marwan A, Simic M, Imad F. Calibration method for articulated industrial robots. Procedia Computer Science 2017.

[8] Zhou S, Song Q, Wang X. A kinematics modeling method of linkage robot based on euler spinning method. In: Xu X, editor. Proceedings of 2017 IEEE International Conference on Unmanned Systems (ICUS): Oct. 27-29, 2017, Beijing, China. Piscataway, NJ: IEEE; 2017, p. 104–109.

[9] Nubiola A, Bonev IA. Absolute calibration of an ABB IRB 1600 robot using a laser tracker. Robotics and Computer-Integrated Manufacturing 2013.

[10] Švaco M, Šekoranja B, Šuligoj F, Jerbić B. Calibration of an Industrial Robot Using a Stereo Vision System. Procedia Engineering 2014.

[11] Zhuang H, Wu X. Membership function modification of fuzzy logic controllers with histogram equalization. IEEE transactions on systems, man, and cybernetics. Part B, Cybernetics a publication of the IEEE Systems, Man, and Cybernetics Society 2001.

[12] Mao Y, Jing F, Liang Z, Fang Z. Positional accuracy analysis of welding robot under mechanism clearance and elastic deformation. In: 12th World Congress on Intelligent Control and Automation (WCICA). Piscataway, NJ: IEEE; 2016, p. 3253–3258.

[13] Shi S, Wu H, Song Y, Handroos H, Li M, Cheng Y et al. Static stiffness modelling of EAST articulated maintenance arm using matrix structural analysis method. Fusion Engineering and Design 2017.

[14] KUKA AG. KUKA System Software 8.3: Operation and Programming Instructions for System Integrators; 2017.

Acknowledgement

The authors of this research would like to acknowledge the support of Daimler AG, Sindelfingen for funding this work and letting us conduct calibration experiments on their robots. The IGF-project 20057 N (HORuS^2) of the research association FVP (Forschungsvereinigung Programmiersprachen für Fertigungseinrichtungen e.V.) was supported via the AiF within the funding program „Industrielle Gemeinschaftsforschung und –entwicklung (IGF)“ by the Federal Ministry of Economic Affairs and Technology (BMWi) due to a decision of the German Parliament.

IOP Conf. Series: Materials Science and Engineering **575** (2019) 012014 doi:10.1088/1757-899X/575/1/012014

Fault Detection and Diagnosis for a Nonlinear Hydraulic Actuator

Juan Felipe Ortiz, Carlos Borrás and Diego Sepulveda

Mechanical Engineering Faculty, Universidad Industrial de Santander, 680003 Bucaramanga, Colombia

juanfortiz60@gmail.com; juan2168908@correo.uis.edu.co

Abstract. In this work a nonlinear dynamic and mathematic model for the position control of a hydraulic actuator with internal leakage was developed, simulations were made and a basic PID control in order to check the actuator behaviour. An internal leakage emulator was installed bypassing the two chambers of the hydraulic cylinder. Time and position data were acquired via Quanser Q8-USB and saved for modelling and validation purposes, sinusoidal and square signals and both for four fault conditions. Data were classified through MATLAB Simulink algorithm having in consideration parity equations and residual generation, obtaining a high level of validation for the proposed model and the desired responses and fault detection, diagnosis and classification.

1. Introduction

Hydraulic actuators are widely used in industrial applications due to their excellent power-size relation, their capacity to apply high forces indefinitely and their high response [1]. Hydraulic servo systems combine the advantages of hydraulic actuators with the versatility of electronics to give the most powerful actuators in modern industrial applications like robotics, aerospace industry, mining, testing equipment and production lines [2] [3].

However, electrohydraulic actuators exhibit a highly and complex nonlinear dynamic performance [4] [5]. The main nonlinearities are due to compressibility of the hydraulic fluid, the flow-pressure relationship in the servo valve, friction forces and internal and external leakage in the actuator [6] [7]. In this work a nonlinear mathematic and dynamic model was first developed having in consideration the internal leakages through the two chambers of the hydraulic actuator. Experimental validation was made in Dynamics and Structural Control Lab at Universidad Industrial de Santander in order to check the accuracy of the dynamic model. Fault detection was made with the residual generated between the two models and fault classification and diagnosis was carried out with features extraction having in consideration the residual and internal leakage flow in order to create boundaries to successfully classify the case studies.

2. Nonlinear Dynamic Model

The schematic representation of the system under consideration is shown in figure 1. The hydraulic actuator is double rod and is commanded by a servo valve. The objective is tracking precisely a specified displacement trajectory.

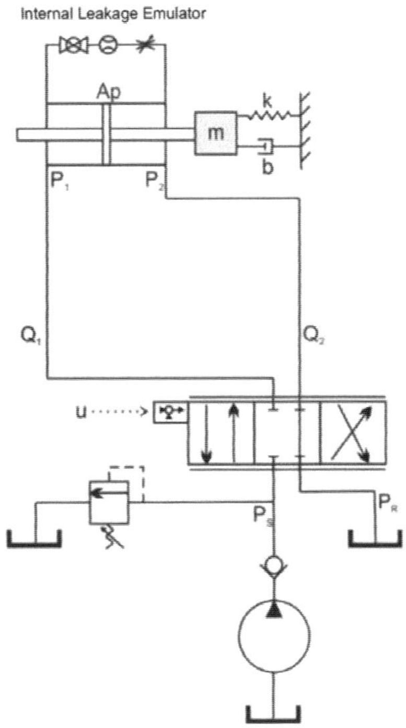

Figure 1. Hydraulic system schematic.

2.1. Actuator Analysis
The motion dynamics of the mass of the actuator is obtained applying Newton's second law:

$$m\ddot{x}_p = A_p P_L - b\dot{x}_p - kx_p \quad (1)$$

where m, b and k are the mass, the viscous damping coefficient and the mass stiffness. x_p is the actuator's displacement, A_p the effective annular piston area and P_L the load pressure, defined as $P_L = P_1 - P_2$.The actuator's dynamics is obtained by applying the continuity principle:

$$Q_L = A_p\dot{x}_p + \frac{V_t}{4\beta}\dot{P}_L \quad (2)$$

where β is the effective bulk modulus, V_t the total fluid volume trapped in the actuator's chambers and system's pipes, Q_{leak} is the leakage flow between the two chambers and Q_L is the flow supplied by the servo valve to move the load, defined as.

$$Q_L = Q_{Leak} + \frac{Q_1 + Q_2}{2} \quad (3)$$

2.2. Servo valve Analysis
The flow supplied by the servo valve is related with the spool displacement by the following equation, assuming that the orifices are symmetric:

$$Q_L = C_d w x_v \sqrt{\frac{P_S - P_L}{\rho}} \quad (4)$$

IOP Conf. Series: Materials Science and Engineering **575** (2019) 012014 doi:10.1088/1757-899X/575/1/012014

where C_d is the discharge coefficient through the orifice, w is the area gradient of the orifice, x_v is the spool displacement, P_S is the supply pressure y ρ is the hydraulic fluid density. The relationship between control voltage and spool displacement is obtained:

$$u = \frac{1}{k_v} x_v \tag{5}$$

$$x_v = uk_{ev} \tag{6}$$

where k_v is the gain of the servo valve, and u the control voltage.

Combining equations, flow equation remains:

$$Q_L = uk_v C_d w \sqrt{\frac{P_S - P_L}{\rho}} \tag{7}$$

2.3. Nonlinear Dynamic Model Without Fault

Using equations 1, 2, 3, 4, 5 and assuming that $Q_{Leak} = 0$, the nonlinear dynamic model without fault are constructed as follows:

$$\ddot{x}_p = \frac{1}{m}\left[A_p P_L - b\dot{x}_p - kx_p\right] \tag{8}$$

$$\dot{P}_L = \frac{4\beta}{V_t}\left[uk_v C_d \omega \sqrt{\frac{P_S - P_L}{\rho}} - A_p \dot{x}_p\right] \tag{9}$$

The nonlinear plant without fault is built using these nonlinear equations 8 and 9 and is shown in Figure 2. The input to the plant is the servo valve voltage u and the output is the actuator's position x_p which completes the feedback look of the system.

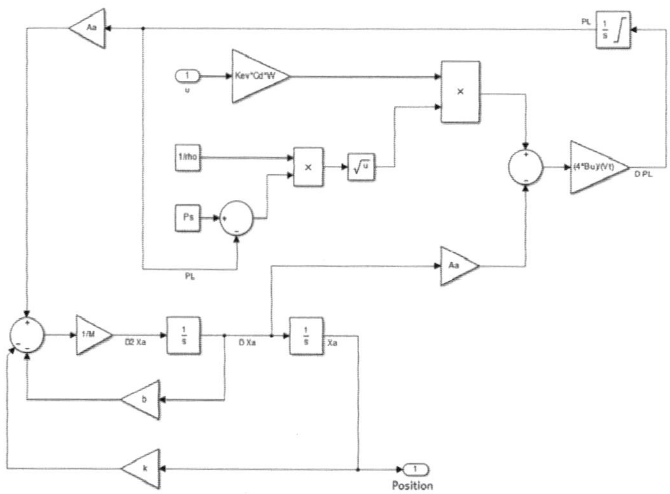

Figure 2. Nonlinear model subplant without fault modelled in Simulink.

2.4. Nonlinear Dynamic Model Without Fault

Using equations 1, 2, 3, 4, 5 and having in consideration the internal leakage inside the actuator, the nonlinear dynamic model with fault are constructed as follows:

$$\dot{P}_L = \frac{4\beta}{V_t}\left[Q_L - Q_{leak} - A_p\dot{x}_p\right] \tag{10}$$

O According to Sepehri [8], the internal leakage flow can be model as:

$$Q_{leak} = C_{leak}w_{leak}\sqrt{\frac{2P_L}{\rho}} \tag{11}$$

where C_{leak} is the flow through a hole coefficient and C_{leak} is the leakage area. Through mathematical operations, the main equations are:

$$\ddot{x}_p = \frac{1}{m}\left[A_pP_L - b\dot{x}_p - kx_p\right] \tag{12}$$

$$\dot{P}_L = \frac{4\beta}{V_t}\left[uk_vC_d\omega\sqrt{\frac{P_S-P_L}{\rho}} - C_{leak}w_{leak}\sqrt{\frac{2P_L}{\rho}} - A_p\dot{x}_p\right] \tag{13}$$

The nonlinear plant with fault is built using the equations 12 and 13 and is shown in Figure 3. The input to the plant is the servo valve voltage u and the output is the actuator's position x_p which completes the feedback look of the system.

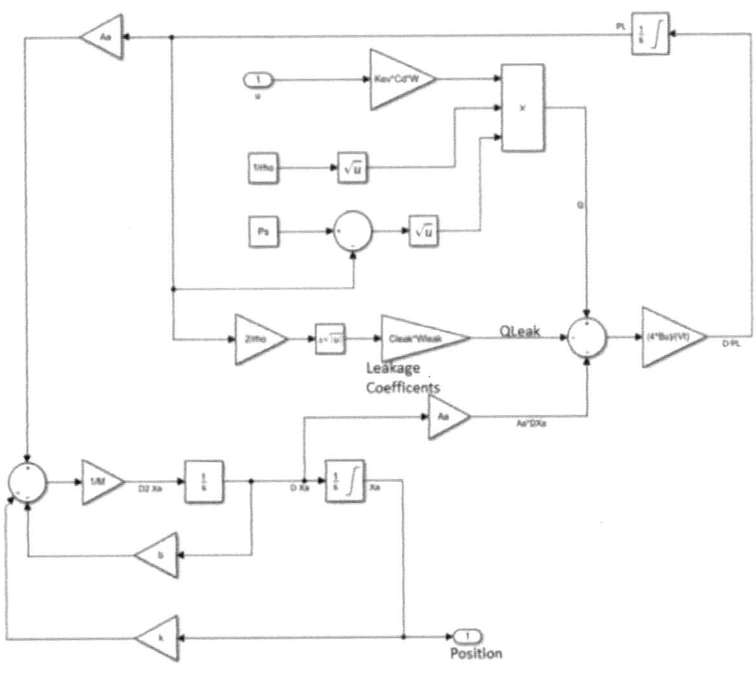

Figure 3. Nonlinear model subplant with fault modelled in Simulink.

3. Simulation

The simulations of the position of the hydraulic actuator were carried out using two types of signal inputs: sine and square wave. For both signals the position set-point was 0.15 [m], which is the half of the actuator's stroke.

Simulations were made through computerized software (MATLAB and Simulink). Faulty and nonfaulty models were analysed and compared using parity equations in order to generate a residual for fault detection, evaluation and classification.

3.1. Normal system simulation

In this case, the Simulink nonlinear dynamic model without fault shown in Figure 2 are used. The physical values and constants were used from a Karpenko and Sepehri [8] research and are listed in Table 1 in order to verify the model behaviour, but in this case, $w_{leak} = 0$

3.2. Faulty system simulation

In this case, the Simulink nonlinear dynamic model with fault shown in Figure 3 are used. The physical values and constants were used from a Karpenko and Sepehri [8] research and are listed in Table 1 in order to verify the model behaviour. In this case, $w_{leak} = 5e^{-7} \frac{mm^2}{mm}$ so that the fault can be noticed.

3.3. Systems comparison

For fault detection and diagnosis, is important to compare the system behaviour in both scenarios. This simulation integrates the two previous models in which flow and position data were compared to generate a residual that is essential for a correct fault detection and diagnosis

The position error or residual are generated by the following equation and Simulink block model were made and represented in Figure 4:

$$r'(s) = [GP(s) - GM(s)]u(s) \quad (14)$$

Table 1. Constants and physical values for modelling without fault.

Parameter	Symbol	Nominal Value	Units
Supply pressure	Ps	17,2	Mpa
Total mass of piston, rods and load	m	12	Kg
Viscous damping coefficient	b	1000	$\frac{N\,s}{m}$
Load spring constant	k	75	$\frac{kN}{m}$
Piston annulus area	A	633	mm^2
Cylinder volume	V	468	mm^3
Leakage orifice coefficient of discharge	Cleak	0,7	--
Leakage orifice area	Wleak	5e-07	$\frac{mm^2}{mm}$
Hydraulic fluid density	ρ	847	$\frac{Kg}{m^3}$
Effective bulk modulus	Bu	689	Mpa
Servo valve coefficient of discharge	Cd	0,6	--
Servo valve orifice area gradient	w	20,75	$\frac{mm^2}{mm}$
Servo valve spool position gain	Kev	0,0406	$\frac{mm}{V}$

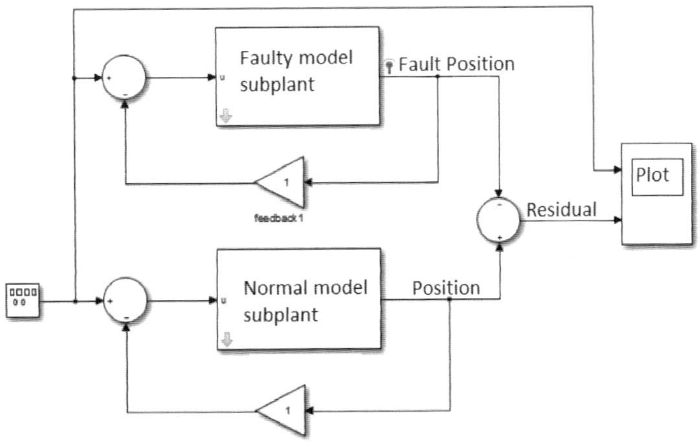

Figure 4. Simulink blocks model comparing the faulty and nonfaulty model for residual generation.

Figure 5 shows the position of both models with a sine signal as input, and the residual generated having in consideration the equation 14.

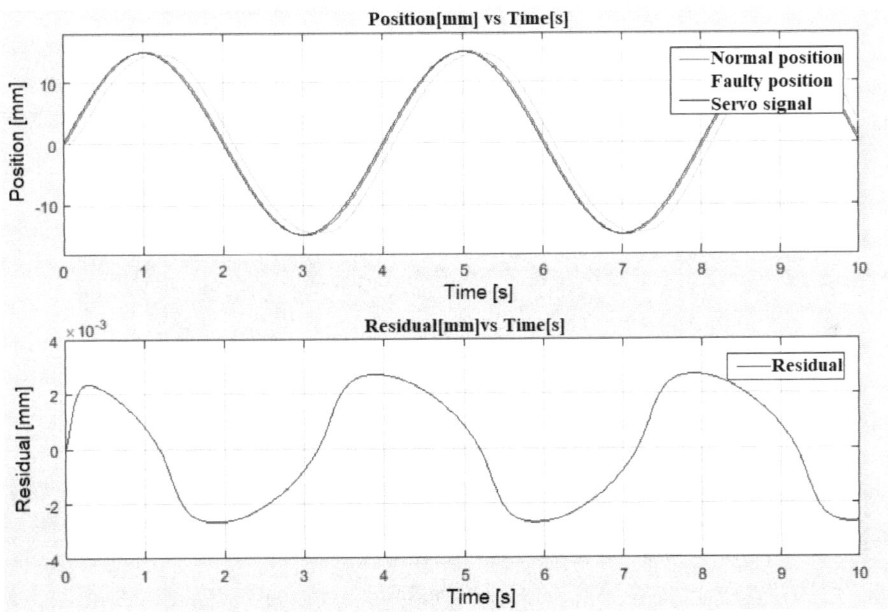

Figure 5. Residual obtained from the faulty and nonfaulty model comparison for sine wave input.

In both cases, the fault is noticeable, the hydraulic actuator can't reach the set point and presents a delay with the input signal. The residual shows a disturbance and error between the two signals and shows an evident fault.

4. Experimental Model Validation

The validation was done in the hydraulic uniaxial seismic shake table of the Dynamics and Structural Control Lab at Universidad Industrial de Santander (shown in Figure 7). The seismic table consists of a Parker hydraulic cylinder, a MOOG 76-263 servo valve, Quanser Q8 USB data acquisition and MTS hydraulic unit and complementary equipment to carry out the assembly of the hydraulic uniaxial seismic shake table shown in Figure 6.

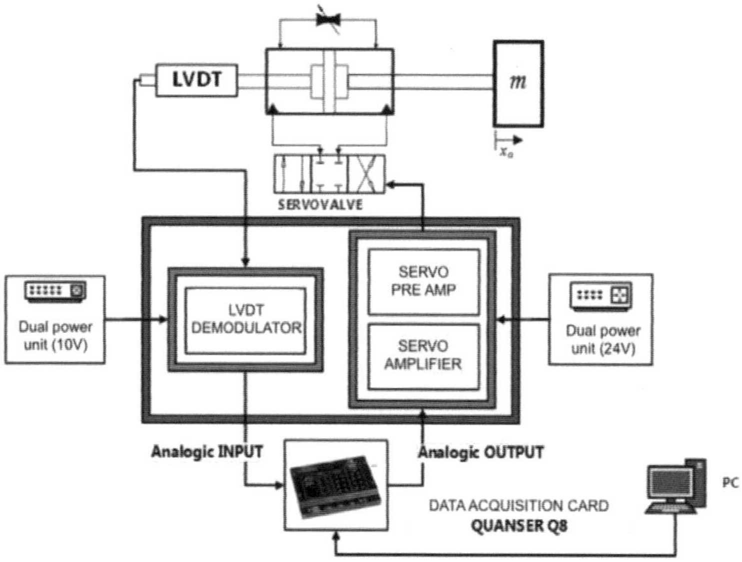

Figure 6. Schematic of hydraulic and electric assembly used to set up the hydraulic uniaxial seismic shake unit and achieve the experimental data acquisition.

Figure 7. Experimental assembly of the hydraulic uniaxial seismic shake table.

4.1. Data acquisition
Data acquisition were made in groups of five vectors, for a 4-inch amplitude and frequency of 0,1Hz, 1Hz and 10Hz both for a square and sinusoidal input signal. The five vectors were averaged to reduce possible errors.

4.2. Validation
The MATLAB and Simulink nonlinear model was calibrated with the real parameters of the experimental assembly with the purpose of have a computational model which can simulate the real assembly with minimal error. These parameters are shown in the Table 2.

Table 2. Constants and physical values for modelling without fault.

Parameter	Symbol	*Nominal Value*	Units
Mass	M	85	Kg
Stroke	Cc	0,1524	m
Active area	Aa	$1{,}62e^{-3}$	m^2
Total volume	Vt	Aa*Cc	m^3
Supply pressure	Ps	$6{,}2e^6$	PSI

For the validation, three different methods were used to confirm the accuracy of the simulation model developed.

The Determination Coefficient: It makes several observations of the variable to be predicted and the variances of the residual and the dependent variable are related.

$$\boldsymbol{\rho^2} = 1 - \frac{\sigma_{res}}{\sigma_{dep}} \tag{15}$$

where σ_{res} is the residual variance and σ_{dep} is the dependent variable variance, in this case the experimental data.

The Correlation Coefficient: It is a measure that aims to quantify the degree of relationship and joint variation between two variables, is a statistical measure that quantifies the linear dependence between two variables.

$$\boldsymbol{R} = \frac{Cov(obs,mod)}{Std(obs)*Std(mod)} \tag{16}$$

where $Cov(obs, mod)$ is the covariance between the model and experimental data, $Std(obs)$ is the standard deviation of experimental data and $Std(mod)$ the standard deviation of the model data.

The Coefficient of Correlation of Concordance: It throws the degree of reproducibility between the data measured in the test bench and the outputs generated by the model.

$$\boldsymbol{CCC} = \frac{A^2+B^2-C^2}{A^2+B^2+D^2} \tag{17}$$

where A and B are the simulation variances and measure data, C is the residual variance and D the average difference between measures.

The absolute error of this case is 0,845%, $\boldsymbol{\rho^2}$ coefficient is 99,94%, $\boldsymbol{R}$ coefficient is 99,97 and $\boldsymbol{CCC}$ coefficient 99,96%. That indicates that the computational model for this case of study describes satisfactorily the real behaviour of the experimental assembly with very small errors.

5. Fault Detection and Classification

Internal leakage was simulated through a needle valve in bypass between the two chambers. The position data were taken in groups of five vectors and for four cases of valve opening (0%, 33,33%, 66,66% and 100%) for a 4-inch amplitude and frequency of 0,1Hz, 1Hz and 10Hz both for a square and sinusoidal input signal. The five vectors were averaged to reduce possible errors.

Figure 8 shows the position of a 0,1Hz sine signal with the four cases of study (without fault, slight fault, medium fault and critical fault).

As expected, the previous image shows that the response of the system varies according to the opening level of the needle valve which, in this specific case, is simulating an internal leakage failure in the actuator.

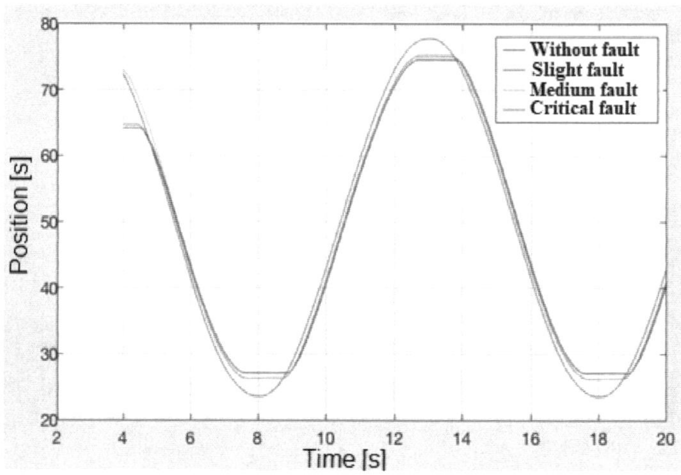

Figure 8. Position data simulation results for 0.1Hz sine wave input with four internal leakage scenarios.

5.1. Classification ranges

According to Watton and Pham [9], A severe failure in the piston seal of a hydraulic cylinder can pass more than 30% of the flow intended by the servo valve for work. Therefore, it will become the point of reference to generate the fault classification ranges.

Using the simulation previously validated with the sample taken, the percentage of opening of the simulated needle valve was manually set so that it would pass 30% of the work flow, as Watton suggests, the system flows are shown in Figure 9.

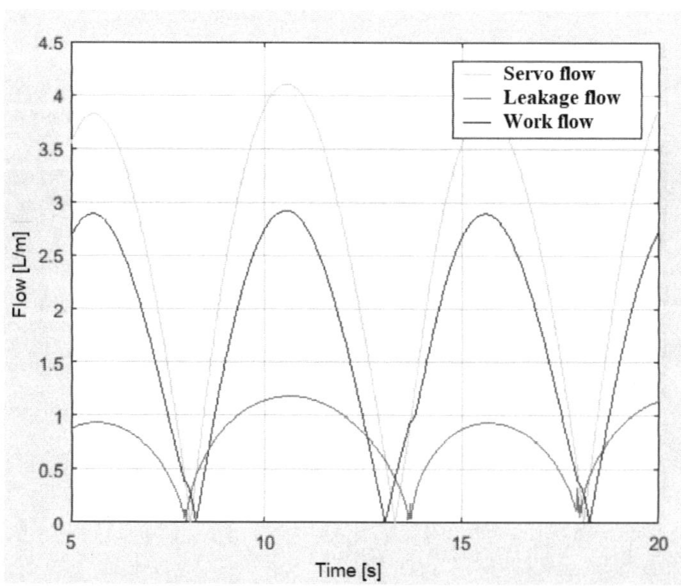

Figure 9. System flows simulation results for 0.1Hz sine wave input when internal leakage is greater than 30% of the work flow.

In addition, the points were searched where 20% and 10% of the flow are lost respectively to classify in three fault levels. Repeating this procedure for the different configurations and having in consideration the fault ranges, Table 3 shows the different flow constants calculated for every scenario.

Table 3. Cleak coefficient for different fault scenarios

Fault	QLeak	0.1 Hz Sqr	0.1 Hz Sin	1 Hz Sqr	1 Hz Sin	10 Hz Sqr	10 Hz Sin
Normal	0	0,1	0,015	0,001	0,001	0,001	0,001
Slight	10%	0,25	0,15	0,35	0,3	0,25	0,12
Medium	20%	0,5	0,35	0,8	0,6	0,7	0,23
Critical	30%	0,85	0,6	1,2	0,9	0,85	0,35
Super	50%	1,4	1,35	2,5	1,9	1,3	0,65

5.2. Fault classification

Considering the classification ranges, a feature extraction of the position signal (Mean, RMS and Standard Deviation) was carried out to improve the accuracy of the classification algorithm hence these values are significantly affected by internal leakage, frequency and type of the signal and they maintain intervals of considerable size, a clear tendency and they remain on the same quadrant.

Fault detection and classification algorithm analyse the residual between the two signals, and compare the flow and features extracted in order to cluster and locate each signal in a classification range generating an output message or alert.

Figure 10 shows how far the cylinder stroke is influenced under the presence of a fault that causes a loss of flow in the piston, from values lower than 10% to 30% and supercritical of 50%.

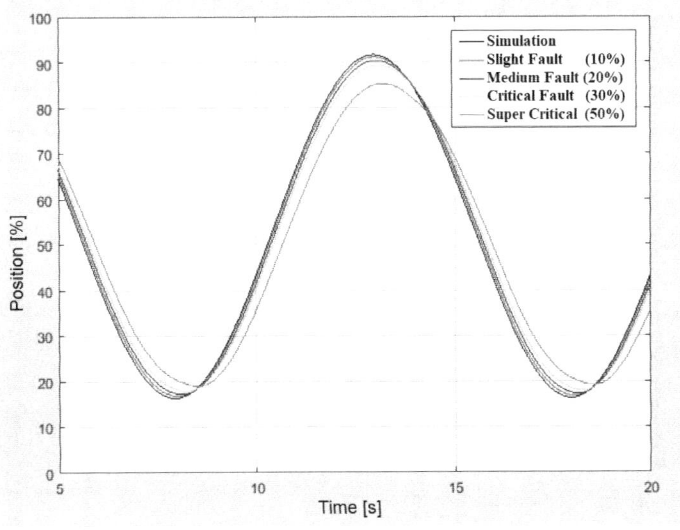

Figure 10. Position simulation results in four fault scenarios (0%, 10%, 20%,30% and 50% of internal leakage) for 0.1Hz sine wave.

Figure 11 shows how the flows behave, according to the simulation, during the test, you can see how the flow sent by the servo valve must be increasing to compensate the loss, but also the work flow is reduced with the increase of the loss flow.

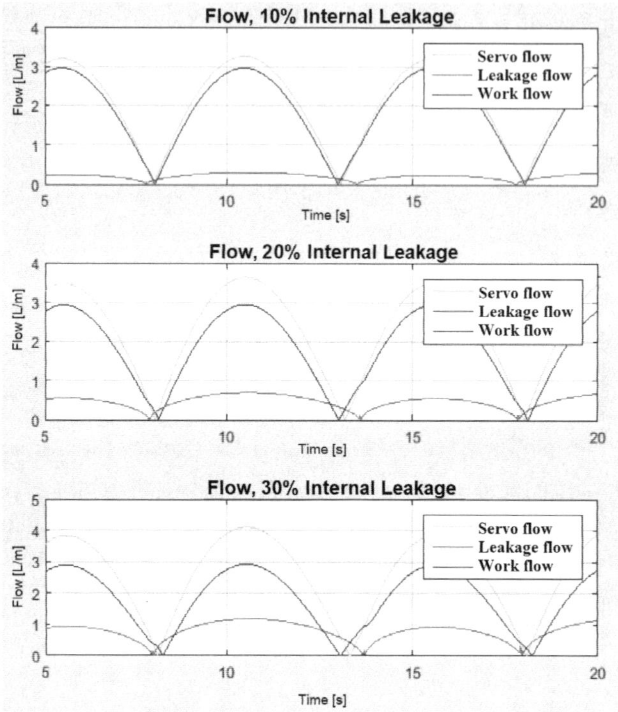

Figure 11. Flows data simulation results for 0.1Hz sine wave input with three internal leakage fault scenarios (10%, 20% and 30% of internal leakage).

Table 4 shows the Leakage coefficient, the flows and the features extracted with the limit values for the clustering and diagnosis.

Table 4. Parameters for fault clustering

Fault	C Leak	Q Servo [L/m]	Q Leak [L/m]	Mean	RMS	Std
Normal	0,015	1,9148	0,0018	0,0215	0,0269	0,0162
Slight	0,15	2,0887	0,2044	0,0508	0,0554	0,022
Medium	0,35	2,3151	0,4709	0,1171	0,1263	0,0472
Critical	0,6	2,5484	0,7797	0,2015	0,2163	0,0788
Super	1,35	3,319	1,6704	0,4186	0,4525	0,172

Having in consideration Table 4, the data previously acquired were passed through the MATLAB fault detection and diagnosis algorithm which process de data and classify successfully all the cases of study (0,1Hz, 1Hz and 10Hz for sine and square wave)

6. Conclusions

- In this work the non-linear dynamic and mathematical model of a hydraulic seismic shake table with and without internal leakage through the two chambers of the actuator was developed.
- Simulations were carried out for sine and square wave as input signal with frequencies of 0,1Hz, 1Hz and 10Hz and validated with experimental data, having correlation coefficients higher than 96% and errors lower than 1%. Fault detection were developed through parity equations and residual analysis of the system.

- Three classification ranges were stablished: Slight fault (0-10% leakage flow) medium fault (10%-20% leakage flow) critical fault (20%-30% leakage flow) and super critical fault (higher than 30% of leakage flow) having in consideration as a several failures the internal leakage higher than 30%.
- Fault diagnosis and classification was realized through MATLAB algorithm which extract important features of the signal and clusters in the four classification ranges for each case of study with 94,4% of accuracy.

7. Observations
Future work will include another fault detection and identification techniques as Artificial Neural Networks and Fault Tolerance.

8. References
[1] F. Basile, P. Chiacchio, and D. Del Grosso. International Journal of Robotics and Automation, 24(1):20–37, 2009
[2] Rolf. Isermann. Fault Diagnosis Aplications. Editorial Springer.350 p.
[3] C. Borrás, J. L. Sarmiento, J. F. Ortiz. ASME, 1 (2018)
[4] Y. Jianyong., Y. GUICHAO, and M. Dawei. 2014. Research Article. School of Mechanical Engineering, Nanjing University of Science and Technology, Nanjing 210094, China. 39.
[5] C. Borrás, J. L. Sarmiento, J. F. Ortiz, XI Congreso Colombiano de Métodos Numéricos, 6, 9 (2017)
[6] Michael G. Skarpetis, Fotis N. Koumboulis, and Achilleas S. Ntellis. 2014 22nd Mediterranean Conference on Control and Automation (MED), (6):1553–1558, 2014.
[7] Hong Ming Chen, Jyh Chyang Renn, and Juhng Perng Su. International Journal of Advanced Manufacturing Technology, 26(1-2):117– 123, 2005
[8] Mark Karpenko and Nariman Sepehri. Journal of Dynamic Systems, Measurement, and Control, 132(5):054505–1 – 054505–7, 2010.
[9] Le, T. T., Watton, J., and Pham, D. T., 1997, Proceedings of The Institute of Mechanical Engineers, Part I, journal of Systems and Control Engineering, 211, pp. 307–317

Acknowledgements
The authors of this work thank the Universidad Industrial de Santander, the Faculty of Physical - Mechanical Engineering and the Mechanical Engineering School for the funding and support.

IOP Conf. Series: Materials Science and Engineering **575** (2019) 012015 doi:10.1088/1757-899X/575/1/012015

Research on Condition Monitoring of Key Components in wind Turbine based on Cointegration Analysis

Xide Sun[1], Dong Xue[2], Runfei Li[3], Xiaogang Li[1], Liye Cui[1], Xishan Zhang[4] and Wei Wu[4]

[1]Jilin Electric Power Technology Development Branch Co., Ltd, 130015 Jilin, China

[2]School of automation engineering, Northeast Electric Power University, 132012 Jilin, China

[3]Hebei Jinglong Sun Equipment Co., Ltd, 055550 Hebei, China

[4]Jilin CPI New Energy Co., Ltd., 130012 Jilin, China

izombiexd@163.com

Abstract. Traditional models based on temperature trend can not comprehensively monitor the condition of wind turbine, besides, the threshold obtained from model can cause fluctuation when wind turbine operates under different conditions. In view of this, this paper presents a novel methodology based on cointegration analysis for continuously monitoring the operating conditions of wind turbine. The first step is to determine the optimal combination of parameters from normal SCADA sample data, by using the cointegration test. Then, perform the method of cointegration analysis to calculate the cointegration residuals and stationary threshold line under normal working space. When wind turbine component is operating out of the normal track, the data sample in cointegration residuals will exceed the stationary threshold line, and send the alert signals. The method is tested using SCADA data of wind turbine with known faults, the results demonstrate the proposed method can effectively monitor the abnormal state of generator and gear box, and provide the function of early warning.

1. Introduction

In recent years, with the development of wind power technology, the number and capacity of wind turbines have increased rapidly. However, due to the adverse operation environment and long time varying load operation, the wind turbines have a high failure rate [1]. Therefore, condition monitoring and fault diagnosis of wind turbines at early stage of fault occurrence is necessary. The most widely used condition monitoring methods are oil monitoring [2] and analysis of current (voltage) signal [3] and so on. These technologies can monitor the operating condition of wind turbine in real time, but there is a certain coupling relationship between each component when wind turbine is operating.

To solve these problem, method of online monitoring based on SCADA data has been gradually developed [4],[5]. This kind of method only needs to process the real-time of SCADA data, and even when the wind speed is changing, the fault information can be accurately judged. Therefore, it has the advantages of high efficiency, low cost and low complexity. At present, the analysis method based on SCADA data has been studied in Ref. [6-7]. These studies focus on two ideas: the first is to determine the fault threshold based on the residuals between the fault and normal parameters, and continuously monitor the condition of wind turbine, but the actual operating conditions of wind turbine are constantly changing, this can cause the fluctuation of threshold. The second is to predict the SCADA fault data by using the intelligent algorithms (such as neural network, SVM, etc.). However, intelligent algorithms are complicated and require massive training data, but it is difficult to satisfy for those wind farms put into operation for a short time.

This paper proposed a novel methodology based on cointegration analysis for continuously monitoring the operating conditions of wind turbine. The vital idea is to perform cointegration analysis between the monitored nonlinear variables. Besides, we can obtain the cointegration residuals, this parameter can represent the long-term stable equilibrium relationship between variables [8]. When the wind turbine runs normally, the relative SCADA parameters will change in the normal working space. On the contrary, they will have abnormal changes, this will cause some points in cointegration residuals exceed the stationary threshold line, and send the alert signals, indicating the occurrence of the failure. Therefore this method can continuously monitor the wind turbine and reliably detect abnormal problems.

2. Introduction and Model Building of Cointegration Analysis

2.1. Concept of cointegration

Let $Y_t = (y_{1t}, y_{2t}, \ldots, y_{nt})^T$ denote an $n \times 1$ vector of nonstationary first order single integer time series. This vector is said to be cointegrated if there exists an $n \times 1$ vector $\beta = (\beta_1, \beta_2, \ldots, \beta_n)^T$ such that:

$$\begin{aligned} \beta^T Y_t &= \beta_1 y_{1t} + \beta_2 y_{2t} + \cdots + \beta_n y_{nt} \\ &=\sim I(0) \end{aligned} \tag{1}$$

Eq. (1) implies that the nonstationary time series in Y_t is cointegrated if there is a linear combination of them that is stationary, the $I(0)$ represents stationary status. This stationary linear combination, denote as $u_{rt} = \beta^T Y_t$, is referred to as a cointegration residual. The vector β is called a cointegration vector.

Let $Y_t = (y_{1t}, y_{2t})^T \sim I(1)$, it is a non-stationary time series with Gauss white noise, and the $I(1)$ represents stationary status. Suppose that Y_t is cointegrated with $\beta = (1, -\beta_2)^T$ .This cointegration relationship is represented as

$$\begin{aligned} y_{1t} &= \beta_2 \sum_{s=1}^{t} \varepsilon_{1s} + \varepsilon_{1t} \\ y_{2t} &= \sum_{s=1}^{t} \varepsilon_{1s} + \varepsilon_{2t} \end{aligned} \tag{2}$$

where the only common stochastic trend is $\sum_{s=1}^{t} \varepsilon_{1s}$.

It shows there is a common trend among nonstationary data, but by using conintegation analysis, we can remove it, and obtain stationary cointegration residual, such that

$$\begin{aligned} u_t &= \beta_t^T Y_t = y_{1t} - \beta_2 y_{2t} \\ &= \beta_2 \sum_{s=1}^{t} \varepsilon_{1s} + \varepsilon_{3t} - \beta_2 \left(\sum_{s=1}^{t} \varepsilon_{1s} + \varepsilon_{2t} \right) \\ &= \varepsilon_{3t} - \beta_2 \varepsilon_{2t} \sim I(0) \end{aligned} \tag{3}$$

2.2. Implementation of proposed method

The condition monitoring of wind turbine cointegration-based are as follows:

- Determine the optimal combination of parameters from normal SCADA sample data, using the cointegration test in the off-line training stage.

- Carry out the method of cointegration analysis to calculate the cointegration residuals and stationary threshold line under normal working space.
- Calculate cointegration residual using the cointegration analysis and SCADA data required from the real-time monitoring of wind turbine.

When the points in cointegration residuals exceed the stationary threshold line, it will send the alert signals.

3. Experimental Verification of Wind Turbine SCADA Data

The SCADA data for wind turbine in this paper is derived from a wind farm of Jilin. The nominal power of wind turbine is 1.5MW, the cut-in wind speed is 3m/s, the SCADA date is acquired at 10-min intervals, during thirty days of September 2017, with two known faults as the generator fault and the gearbox fault. Six days data when wind turbine is operating under abnormal operating conditions were selected from them, and another six days normal data as the train dataset. After processing (eliminate the data when wind turbine shut down and those singular data), we can get two group datasets, each size of it is 963.

3.1. Selection of sample characteristic parameters

A vital parameter in wind energy systems is wind speed, it is closely related to other parameters. By observing the tendency changes between wind speed and other parameters, we can find out those key parameters, and the selected parameters are shown in the fig 1. It presents there is a nonlinear relation between the generator speed, the generated power, the generator temperature, the gearbox temperature and wind speed. Thus, those parameters can be used as the sample characteristic parameters.

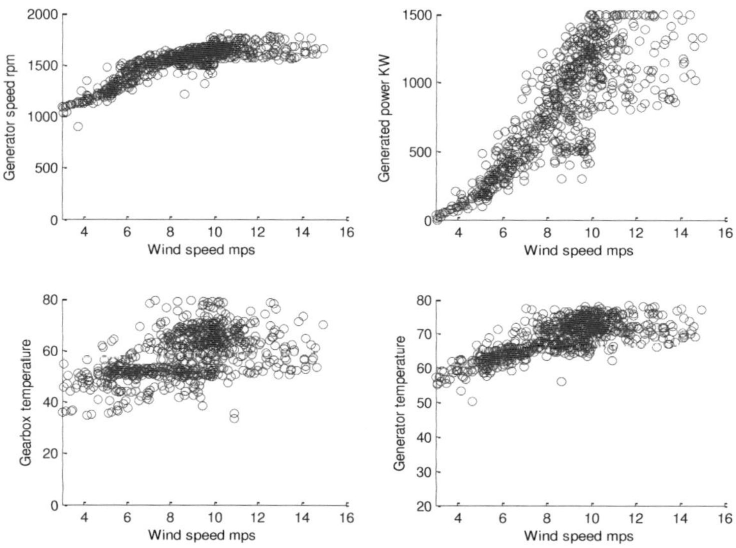

Figure 1. Nonlinear relationship between characteristic parameters and wind speed.

3.2 Analysis of SCADA fault data

First, we do some simple analysis to the fault sample data. Tendency changes of sample parameters are displayed in fig 2. Including the parameter, the generated power, is varying between 0MW to 1.5MW, which represents the dataset is obtained from different operating conditions of wind turbine. The two faults, in the data sample 84 and 394, are enlarged locally as described in fig 3 and fig 4.

In fig 3, around the data sample 84, this abnormal operation happened while wind speed started to fall from 8m/s, then stayed around 3-4m/s. Consequently, the generator speed dropped down from 1300 rpm to almost stationary state, and then suddenly rised to nearly 900 rpm, but during this period, the generator power is consistent with the wind speed. In fig 4, around the data sample 394, the second fault happened while the wind speed was relatively stable around 5m/s, but the generator speed and the generator power were suddenly dropped to 0 value, in the same time the gearbox temperature was boosted up to 70°.

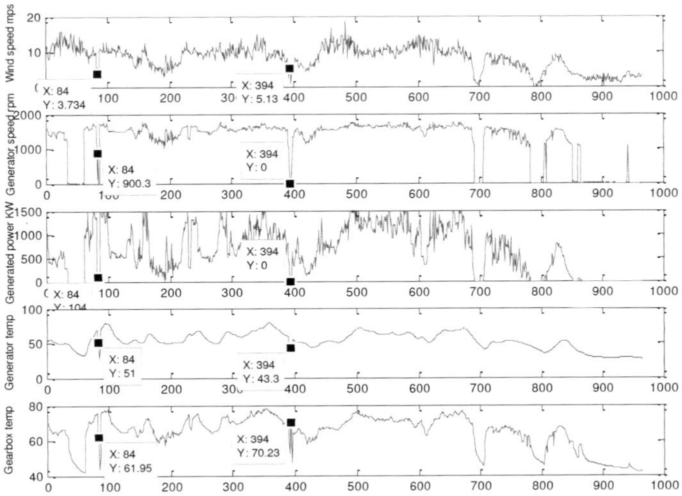

Figure 2. Tendency change of SCADA parameters.

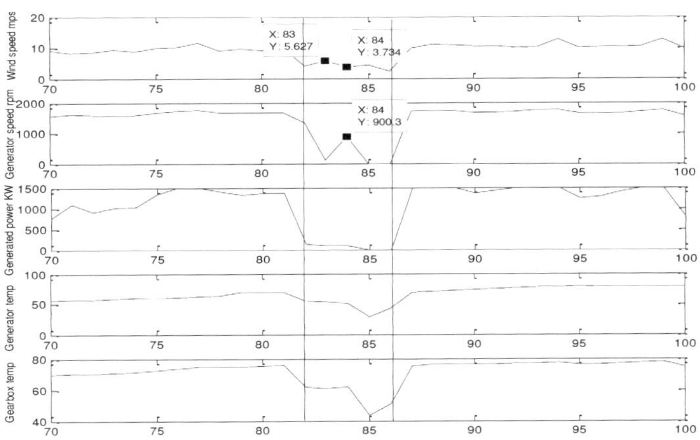

Figure 3. Zoomed data of generator fault.

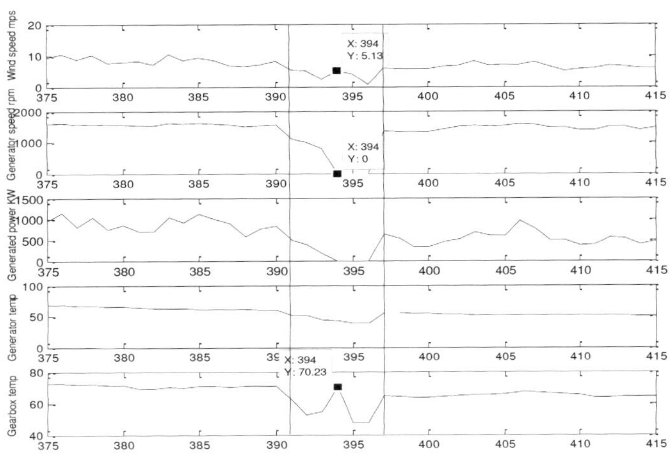

Figure 4. Zoomed data of gearbox fault.

4. Condition Monitoring of Key Components in Wind Turbine

Cointegration analysis can construct stationary relations between nonstationary parameters, when the wind turbine operates abnormally, cointegration residuals will exceed the stationary threshold line, implying a developing fault.

4.1 Cointegration test

Cointegration test is used to find out the cointegration relationships between parameters, those parameters which passed the test can be used to monitor the operating conditions of relevant components. The economics software EViews 8 is used to perform the test by using the common test method, Johannsen cointegration verification (JJ), the experimental object is six days normal dataset. The test results, as shown in table 1, are parameter combinations which has passed the test, the confidence level there is 95%, namely when the test probability is less than 0.05, the hypothetical conditions are invalid.

Table 1. Results after performing cointegration test.

Combinations of SCADA parameters	Cointegration (prob)	Cointegration Vector	Stationary (prob)
Wind Speed Generator Speed Generated Power	0.1405	0.305,1.65E-5,0.0 0254	0.0006
Wind Speed Generator Temp Gearbox Temp	0.0058	0.00059, -0.121,0.243	0.0024
Wind Speed Generator Speed Generator Temp	0.0229	-0.141, -0.0032,0.167	0.0001
Wind S Generator S Generated P Gearbox Temp	0.0075	-0.001,-0.37,0.57, 0.08	0.0007

4.2 Condition monitoring and fault detection using cointegration residuals

The cointegration residuals can be calculated via the cointegration vector which is obtained from last section, the six days fault data were used to validate. The experimental results are shown in fig 5, the horizontal lines indicate the stationary threshold lines which were calculated as $\nu \pm 3\sigma$(the 95% statistical confidence levels), where ν is the mean, and σ is the standard deviation, abnormal problems happen whenever the cointeration residuals go beyond them.

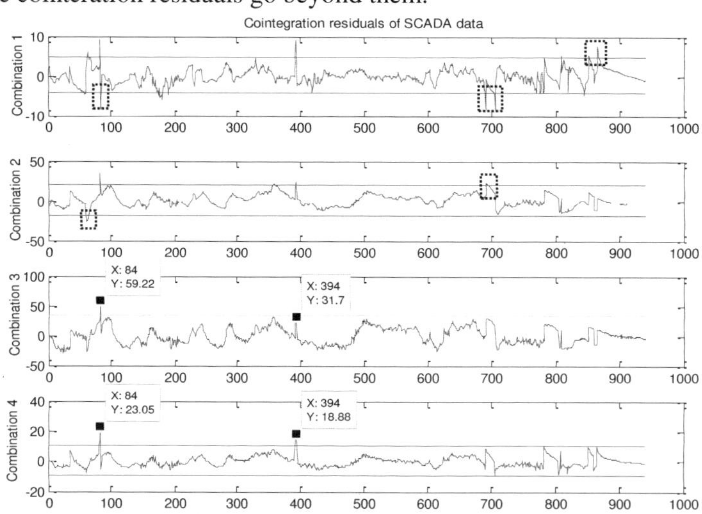

Figure 5. Cointegration residuals of combinations (fault).

However, the first three combinations can not reliably detect the two faults, and eventually cause some misinformation, as the rectangle frames mark. In the contrary, the 4th combination can reliably detect two faults without any mistakes, thus, there select the combination 4, generator speed, the generated power, the gearbox temperature and wind speed as the optimal combination, and its cointegration residuals is displayed in fig 6 (normal and fault).

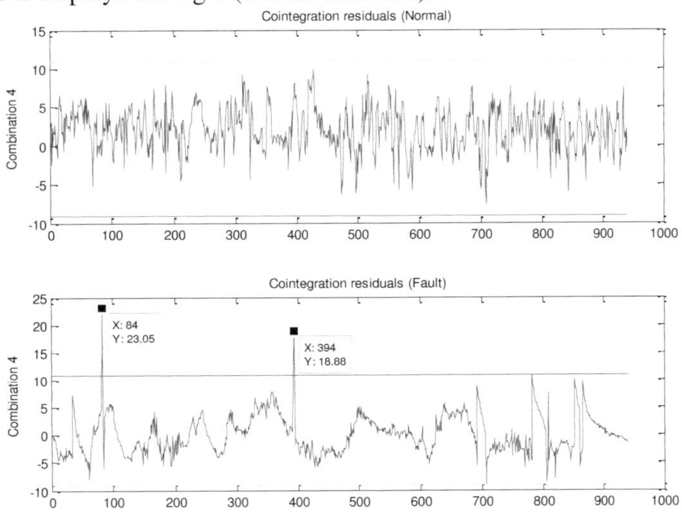

Figure 6. Cointegration residuals of combination 4.

In order to illustrate more specifically, the data sample 84 and 394 of the two faults are enlarged locally is shown in fig 7. It reveals the fault 1, the generator fault, can be detected around the data sample 82, 2 data sample (20min) before the occurrence; the fault 2, the gearbox fault, can be detected around the data sample 392.6, more than 1 data sample (14min) before the fault occurrence.

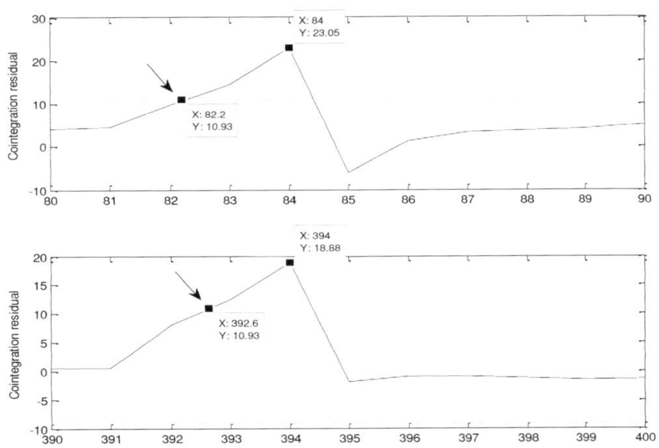

Figure 7. Zoomed data of cointegration residuals.

5. Conclusions

Condition monitoring of key components in wind turbine based on SCADA data can reduce the cost of hardware installation, and diagnose the developing failures in time, especially important for those offshore wind farms which are hard to set up hardware facilities. The proposed method based on cointegration analysis can monitor the key components in wind turbine more comprehensively, avoids the disadvantage of traditional model which can only monitor single parameter. Besides, compared with the current AI technology, this methodology is promising candidate for automated fault diagnosis because of their low complexity, low computation, and cost-efficient. The results demonstrate the proposed method can effectively monitor the abnormal state of generator and gearbox, and provide the function of early warning.

6. References

[1] Mani H and Dahiya R 2017 Condition monitoring of wind turbine for rotor fault detection under non stationary conditions. Ain Shams Engineering Journal, 1 (1) 1-12 (2017)

[2] Fenf Y H, Qiu Y N, Crabtree C J, Long H and Tavner P J 2011 Use of SCADA and CMS signals for failure detection and diagnosis of a wind turbine gearbox[C] //Europe’ s Premier Wind Energy Event, March 14-17, 2011, Brussels, Belgium.

[3] Mohanty A R and Kar C 2006 Fault Detection in a Multistage Gearbox by Demodulation of Motor Current Waveform. IEEE Transactions on Industrial Electronics, 53(4):1285-1297 (2006).

[4] Qiu Y, Feng Y, Tavner P, Richardson P, Erdos G and Chen B 2012 Wind turbine SCADA alarm analysis for improving reliability. Wind Energy, 15 (8) 951-966 (2012).

[5] Yang W, Court R and Jiang J 2013 Wind turbine condition monitoring by the approach of SCADA data analysis. Renew. Energy, 53 365-376 (2013).

[6] Zaher A, McArthur S D J, Infield D G and Patel Y 2009 Online wind turbine fault detection through automated SCADA data analysis. Wind Energy, 12 (6) 574-593 (2009).

[7] Brabanter K D, Karsmakers P and Brabanter J D 2012 Confidence bands for least squares support vector machine classifiers: A regression approach. Pattern Recognition, 45(6): 2280-2287 (2012).

[8] Cross E J, Worden K and Chen Q 2011 Cointegration: a novel approach for the removal of environmental trends in structural health monitoring data, Proc. R.Soc. A 467 (2011) 2712-2732.

Acknowledgements

The research work was supported by Science & Technology Development Plan of Jilin province -Major Science & Technology Bidding under Grant 20180201004SF.

An Intelligent Adaptive Robust Controller Based on Neural Network for a Pneumatic System

Ye Chen[1], Guoliang Tao[1,3], Fangan Meng[2]

[1]State Key Laboratory of Fluid Power & Mechatronic Systems, Zhejiang University, Hangzhou, Zhejiang, China
[2]DJI-Innovations, Shanghai, China

[3]gltao@zju.edu.cn

Abstract. Pneumatic actuator systems are widely applied in industrial automation. This paper proposes an intelligent adaptive robust controller(ARC) based on neural network(NN) to improve the applicability in a pneumatic servo system that consists of a pneumatic cylinder which is controlled by a proportional directional control valve. In this design, adaptive robust controller is provided with suitable control parameters automatically by a real-time neural network depending on feedback errors. So the ARC the tuning process that is time consuming and restricts the application of pneumatic systems is avoided. The experimental results show all the control parameters of ARC are converged and good control performance is achieved.

1. Introduction

Pneumatic actuator systems are of interest for industrial applications as they have the main advantages of high power to weight ratio, cleanliness, simplicity of operation, and low cost [1]. But due to dynamitic highly nonlinear, severe parametric uncertainties and uncertain nonlinearities of pneumatic systems, it is difficult to control the pneumatic systems [2-3].

A large body of research is aimed for improving performance of servo pneumatic systems. For example, conventional linear controllers is developed including PID gain scheduling techniques [4] and nonlinear state feedback techniques [5-6]. As an example of nonlinear control, sliding mode control for pneumatic system is proposed by Ning and Bone [7]. As a more recent example, Deyuan Meng [8] applied ARC to the servo pneumatic system and experiment results showed good tracking performance. ARC is effective as it reduces the extent of parametric uncertainties through utilizing on-line recursive least squares estimation, and attenuates the effects of parameter estimation errors, disturbances and unmodeled dynamics by employing sliding mode control method [9]. However, the tuning process of ARC by try-and-error for pneumatic systems restricts the application of the control scheme.

Neural network has been highly applied for many different areas [10-12]. As the tuning problem, Gi et al. designed feedback linearization controller with a neural network which is training off-line [13]. Results showed tracking performance with NN much better than non-NN-compensated controller. Behrad [14] reported success with a novel NNC with PID for pneumatic gantry robot. Kothapalli and Hassan [15] used an off-line NN to adjust the gains of a PI position controller of a pneumatic system. Simulation shows NN could improve the performance.

This paper sets out to document a novel neural network based intelligent adaptive robust controller to improve the applicability in servo pneumatic actuator. The structure is organized as following.

Section 2 describes the modelling of pneumatic system and the design of ARC briefly. Section 3 explains the details of the design of NN including structure, training process of neural net and the NN based ARC method. Experiments are carried out in Section 4. Finally, Section 5 concludes the paper.

2. Adaptive Robust Controller Design

2.1. Dynamic models

In this paper, the pneumatic servo system mainly consists of proportional directional control valve (FESTO MPYE-5-1/4-010B) and a cylinder (FESTO DNC-63-200-P) and Fig.1 shows the schematic diagram of the servo pneumatic system. The whole mathematical model for the pneumatic servo actuator system includes the model of the pneumatic proportional directional control valve which demonstrate the relation between the output voltage and mass flow rate, then model thermodynamic changes in pneumatic cylinder and motion of the whole valve and cylinder system. This paper only introduces the modelling process briefly, for the details of model is demonstrated in [16-18]. The system model in state-space form is written as (1).

$$\begin{cases} \dot{x}_1 = x_2, \\ m\dot{x}_2 = \bar{A}(x_3 - x_4) - \theta_1 x_2 - \theta_2 S_f(x_2) + \theta_3 + \bar{f}_0, \\ \dot{x}_3 = \dfrac{\gamma R}{S_p V_a}(\dot{m}_{ain} T_s - \dot{m}_{aout} T_a) - \dfrac{\gamma A}{V_a} x_2 x_3 + \dfrac{\gamma - 1}{S_p V_a}\dot{Q}_a + \theta_4 + \tilde{d}_{a0}, \\ \dot{x}_4 = \dfrac{\gamma R}{S_p V_b}(\dot{m}_{bin} T_s - \dot{m}_{bout} T_b) - \dfrac{\gamma A}{V_b} x_2 x_4 + \dfrac{\gamma - 1}{S_p V_b}\dot{Q}_b + \theta_5 + \tilde{d}_{b0}. \end{cases} \tag{1}$$

where $\vec{\theta} = [\theta_1, \theta_2, \theta_3, \theta_4, \theta_5]^T$ are the unknown parameters and $\vec{X} = [x_1, x_2, x_3, x_4]^T$ are the states.

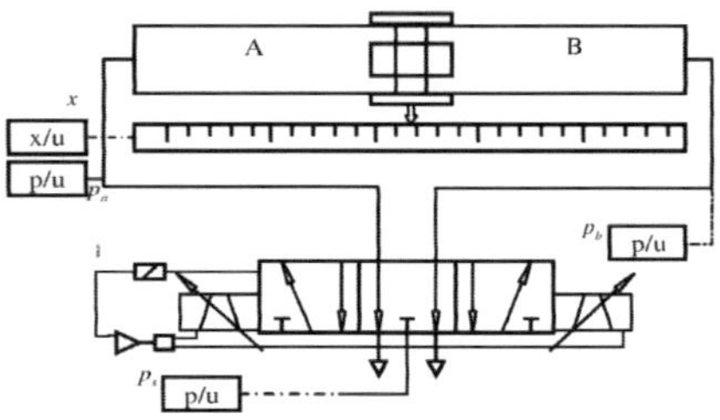

Figure 1. Schematic diagram of servo pneumatic system

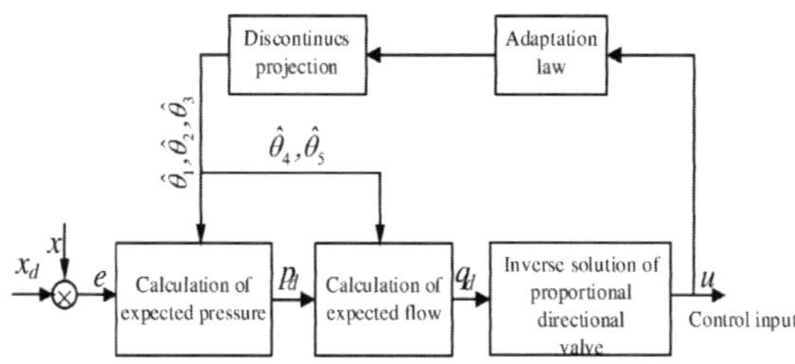

Figure 2. Adaptive robust control schematic for the system

2.2. Design of ARC

The Figure.2 shows the adaptive robust control schematic for the pneumatic system. As shown in Figure.2, the design of ARC is divided into robust control law with backstepping method and adaptation of system parameters. This paper will introduce this part briefly and the details design process is illustrated in [18].

Robust control law design as follows. The Step 1, the expected virtual input pressure p_d is calculated according to the motion equation of piston that is the first and second equation of (1), semi-definite Lyapunov function and feedback information of position sensors. Step 2, the expected virtual control input q_d is calculated according to the mass flow rate function that is the last two equations of (1), semi-definite Lyapunov function and feedback information of pressure sensors. Step 3, q is already obtained, the control input signal u for the proportional directional control valve could be calculated according to mass flow equation and the relation between the input signal valve and area of valve orifice.

As for the adaptation of system unknown parameters, on-line recursive least squares estimation (RLSE) of parameter θ is employed. The RLSE algorithm is limited by projection mapping guaranteed the parameter estimates keep within the closure of the convex set Ω_θ that is a known bound. Using the preset adaption rate restricts for a controlled estimation process, the estimated parameters $\hat{\theta}$ are updated along with the discontinuous projection calculated by RLSE algorithm [8].

3. An intelligent adaptive robust controller

The regulation of control parameters for servo pneumatic system ARC is quiet time consuming and restricts its application. In order to increase the applicability of ARC, a NN based intelligent ARC is proposed. In this approach, NN adapt ARC control parameters online according to feedback errors automatically.

3.1. RBF Neural Network

The Neural Network regulation algorithm is based on the modified back propagation method (MBPM) originated by Lewis[19]. As showed in Figure.3, the basic structure for the NN in that is a three layer neural network. The activation functions for each neuron is a key design deciding the nature of neurons. For radial basic (RBF) NN's, R_j called activation function is commonly given as:

$$R_j(\boldsymbol{x}) = \exp\left(-\frac{\left\|\boldsymbol{x}-\boldsymbol{c}_j\right\|^2}{b_j^2}\right),\ j=1,2,\dots N \tag{2}$$

where c_j is the center of j th activation function, b_j is the width of jth activation function, N is the number of nodes in second layer. Input of neural network is defined as a n dimensional vector $\boldsymbol{x}=[x_1(t),x_2(t),\dots,x_n(t)]^T$. And the output is given as:

$$y_m(t) = \sum_{j=1}^{N} \omega_j(t) R_j(\boldsymbol{x}(t)) + d \tag{3}$$

where ω_j is the weight of j th element of second layer and d is bias of second layer.

The target function for Neural Network practice is

$$E(t) = \sum_{i=1}^{M} e_i^2(t) = \sum_{i=1}^{M} y_i(t) - y_{im}(t) = y_i(t) - \sum_{j=1}^{N} \omega_j(t) R_j(t) - d \tag{4}$$

where $e_i(t)$ is the error and $y_i(t)$ is the output of real system, M is the number of data considered for calculation for the system.

To express in matrix form, we define $y=[y_1,y_2...y_M]^T$, $u=[R_1,R_2...R_N,1]^T$, $U=[u_1^T,u_2^T...u_M^T]^T$, $\omega=[\omega_1,\omega_2...\omega_N,d]^T$. The target function becomes

$$F(\omega) = (y-U\omega)^T(y-U\omega) \tag{5}$$

Let regularization and expand the expression can get general form of a quadratic function and the Hessian matrix as $2[U^TU+\rho I]$ which is semi-positive or positive [20].

For the regulation of parameters, matrix is supposed to be positive, then the strong minimum is:

$$\omega = [U^TU+\rho I]^{-1}U^Ty \tag{6}$$

3.2. Parameter selection of activation function

The parameters of neurons affect the effective of neural network greatly. This part explains the selection and computation of neural network.

The parameter c which means the center of activation function and parameter b decided the width of activation function play an important role in training process of neural network. According to [20-21], this paper pick the parameter c according to the orthogonal least squares method that means centers focus on the input data of neural network and b is chosen as $b=1/\sqrt{2}dist$ where $dist$ is the average distance between its centers and neighbors.

For the number of neural N and the number data M take in for calculation, a trade-off is made between the precision of neural network and calculation efficiency. The Larger the N, M is, the neural network is more precision for the target function at cost of large computing time. In this paper we set N=6, M=8, satisfying the demand of quick computing which is in real time and guaranteed the relative precision.

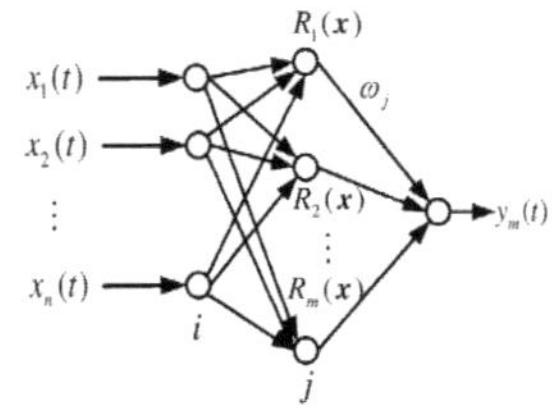

Figure 3. RBF Neural Network

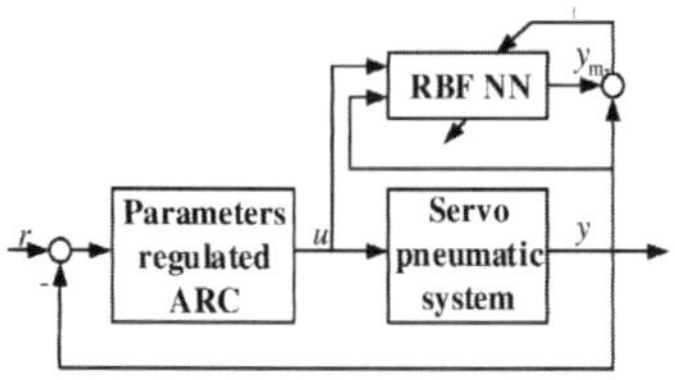

Figure 4. Schematic structure of NN based ARC

3.3. NN based ARC

In this section, NN based ARC strategy is proposed and the schematic structure of arithmetic is showed in Figure.4. Define control input is *u*, control parameter is *p*, e_c is the error between system reference trajectory *r* and system output *y*,

$$\begin{cases} x_c(t) = e_c(t) - e_c(t-1) \\ \Delta u(t) = p(t-1)x_c(t) \end{cases} \tag{7}$$

The target function is as (4) and gradient decent method is used for the adaptation process of parameter *p* as,

$$\Delta p(t) = -\eta_c \frac{\partial E(t)}{\partial p(t-1)} = -\eta_c \frac{\partial E(t)}{\partial y(t)} \frac{\partial y(t)}{\partial u(t)} \frac{\partial u(t)}{\partial p(t-1)} = \eta_c e_c(t) \frac{\partial y(t)}{\partial u(t)} x_c(t) \tag{8}$$

and learning of parameter *p* is,

$$p(t) = p(t-1) + \Delta p(t) + \alpha_c(p(t-1) - p(t-2)) \tag{9}$$

According to (8), $\frac{\partial y(t)}{\partial u(t)}$ (Jacobian matrix) need to be calculated when adapt parameters. Because Jacobian matrix is unknown using RBF NN output instead of real system output, that is

$$\frac{\partial y(t)}{\partial u(t)} \approx \frac{\partial y_m(t)}{\partial u(t)} = \sum_{j=1}^{m} \omega_j(t-1) \frac{\partial R_j(\boldsymbol{x}(t))}{\partial u(t)} = \sum_{j=1}^{m} \omega_j(t-1) \partial R_j(\boldsymbol{x}(t)) \frac{c_{j(n_y+1)}(t-1) - u(t)}{b_j^2(t-1)} \tag{10}$$

For the pneumatic system, the control parameters and the state of system is described as the following nonlinear extended autoregressive sliding mode,

$$y(t) = f(y(t-1), \ldots, y(t-n_y); u(t-1), \ldots u(t-n_u)) \tag{11}$$

where $u(\bullet)$, $y(\bullet)$ represent output and input of system, and $f(\bullet)$ represents the relation between output and input. k_1, k_2, k_3 is the control parameters need to regulate and the adaption of control parameters is on the basis of relation between control parameters and the error defined in ARC as follows,

$$\begin{cases} \Delta k_1(t) = p_{11}(t-1) \times x_{1cm} \\ \Delta k_2(t) = p_{21}(t-1) \times x_{1cm} + p_{22}(t-1) \times x_{2cm} \\ \Delta k_3(t) = p_{31}(t-1) \times x_{1cm} + p_{32}(t-1) \times x_{2cm} \\ + p_{33}(t-1) \times x_{3cm} \end{cases} \tag{12}$$

where $x_{icm} = e_{icm}(t) - e_{icm}(t-1)$, $x_{icrms} = e_{icrms}(t) - e_{icrms}(t-1)$, *i*=1,2,3, $e_{cm} = \frac{1}{10}\max(10)\{|e_c|\}_{0 \le t \le T}$ is the average value of 10 largest error of during a period of tracking signal.

Taking k_3 as an example, the adaptation of control parameter is

$$\begin{cases} \Delta p_{31}(t) = -\eta_c \frac{\partial E_c(t)}{\partial k_3(t-1)} = -\eta_c \frac{\partial E_c(t)}{\partial y(t)} \frac{\partial y(t)}{\partial k_3(t)} \frac{\partial k_3(t)}{\partial p_{31}(t-1)} = -\eta_c e_c(t) \frac{\partial y(t)}{\partial k_3(t)} x_{1cm} \\ \Delta p_{32}(t) = -\eta_c \frac{\partial E_c(t)}{\partial k_3(t-1)} = -\eta_c \frac{\partial E_c(t)}{\partial y(t)} \frac{\partial y(t)}{\partial k_3(t)} \frac{\partial k_3(t)}{\partial p_{32}(t-1)} = -\eta_c e_c(t) \frac{\partial y(t)}{\partial k_3(t)} x_{2cm} \\ \Delta p_{33}(t) = -\eta_c \frac{\partial E_c(t)}{\partial k_3(t-1)} = -\eta_c \frac{\partial E_c(t)}{\partial y(t)} \frac{\partial y(t)}{\partial k_3(t)} \frac{\partial k_3(t)}{\partial p_{33}(t-1)} = -\eta_c e_c(t) \frac{\partial y(t)}{\partial k_3(t)} x_{3cm} \end{cases} \tag{13}$$

Despite the coupling of control parameters, the 3 control parameters k_1, k_2, k_3 which influence the control precision greatly will be regulated simultaneously with the designed NN.

4. Experiment

4.1. Experiment setup

To verify the NN based ARC, an experimental platform is set up. FESTO DNC-63-200-P cylinder is controlled through a FESTO MPYE-5-1/4-010B proportional directional valve. MPX5700AP pressure transducers are used for the measurement of the chamber pressures and the tank pressure. Position of the cylinder movement is acquired by the resistance displacement sensor (KTC-600). Velocity and acceleration are calculated by differential tracking filter. NI PXI 1042Q controller board is used for providing control output.

4.2. Trajectory tracking experiment

The initial value of NN algorithm parameters and ARC control parameters are shown in table 1.

Table 1. Design parameters of intelligent control strategy

symbol	Meaning	value
$b_j(0)$	Initial value activation function parameter	1
$c_j(0)$	Activation function parameter	Calculated
$\omega_j(0)$	Activation function parameter	Calculated
N	Number of hide layer nodes	6
M	Data for calculation for one time	8
α_c	Momentum factor of regulating parameters	0.1
η_c	Learning rate of regulating parameters	0.5
k_1,k_2,k_3	Initial value of control parameters k_1,k_2,k_3	$[1,1,1]^T$

Figure.5 provides the convergence of parameters k_1,k_2,k_3 and e_1,e_2,e_3 for NN based Intelligent ARC tracking 0.5Hz sinusoidal wave. It is showed that the parameters k_1,k_2,k_3 converge to a steady value in a short time automatically and error e_1,e_2,e_3 reduce gradually. Figure.6 shows tracking performance when all the control parameters are steady.

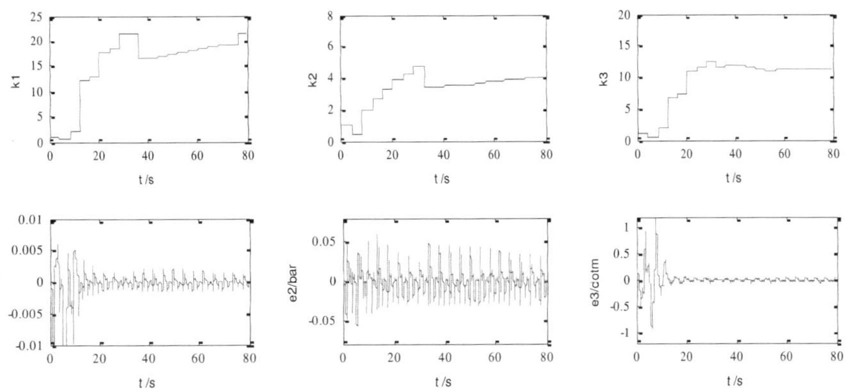

Figure 5. Convergence of k_1,k_2,k_3 and e_1,e_2,e_3

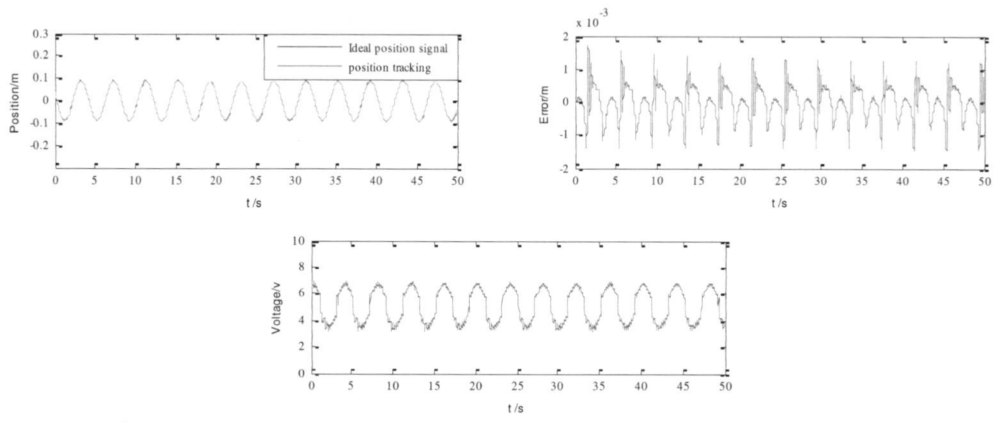

Figure.6 Tracking responses of 0.09sin($0.5\pi t$) for steady

The experiment can be qualified with the following two indices:

1) $e_m = \max\{|z_1|\}_{T_f-10\le t\le T_f}$, the maximum absolute tracking error in last 10 seconds, is measured the final tracking accuracy, where T_f represents the total running time.

2) $\|e\|_{rms} = \sqrt{\frac{1}{10}\int_{Tf-10}^{Tf} z_1^2 dt}$, the root-mean-square of tracking error, is measured the average tracking accuracy.

5. Conclusion

1) An Intelligent ARC based NN is developed for both strengthening the applicability of pneumatic cylinders and achieving high precision tracking control. Experimental results for sinusoidal trajectory tracking motion are shown in Figure.5 and Figure.6. The experiment illustrates that all three control parameters k_1, k_2, k_3 are converged simultaneously despite the existence of fluctuation caused by the coupling of parameters and disturbance, and good tracking performance is achieved that the relative maximum tracking error is less than 1.5% and root-mean-square error is less than 0.5%.

2) The proposed controller employs on-line NN to regulate the control parameters of ARC automatically with no harm to the control performance of ARC in real time. The method points out how to design a neural network including selection of neural number, calculation of center, width, bias and weight of neural and how to adapt parameters. This method is not constraint of utilizing of ARC for regulation of control parameters if good target function is selected.

3) There are several difficulties to overcome. The valve deadzone properties which restrict the control precision is a vital problem for application of pneumatic system. Even the same batch of valves has different deadzones, but it is unrealistic to measure each one's deadzone. A method is need to propose to solve this problem. Also friction of pneumatic system should be considered for a higher precision control. A new model or an intelligent method compensate the friction, such as neural network which has been proved effectively for fitting complex model need to be proposed.

6. References

[1] TAGHIZADEH M, NAJAFI F, GHAFFARI A. Multi-model PD-control of a pneumatic actuator under variable load [J].International Journal of Advanced Manufacturing Technology, 2010,48(5/8): 655-662.

[2] Z. Šitum, D. Kosić, and M. Essert, "Nonlinear mathematical model of a servo pneumatic system," 9th International research / Expert Conference, TMT, Antalya: Turkey, 26-30 Sept 2005.

[3] E. Richer and Y. Hurmuzlu, "A high performance force actuator system Part-2: Nonlinear controller design," ASME Journal of Dynamic Systems Measurement and Control, vol. 122, no. 3, Sept. 2000, pp. 426-434.

[4] SITUM Z, PAVKOVIC D, NOVAKOVIC B. Servo pneumatic position control using fuzzy PID gain scheduling [J]. J Dyn Syst, Meas, Control, 2004, 126(2): 376-387.
[5] XIANG F, WIKANDER J. Block-oriented approximate feedback linearization for control of pneumatic actuator systems [J]. Control Engineering Practice, 2004, 12: 387-399.
[6] SCHULTE H, HAHN H. Fuzzy state feedback gain scheduling control of servo-pneumatic actuators [J]. Control Engineering Practice,2004, 12: 639-650.
[7] Gary M. Bone,Shu Ning. Experimental Comparison of Position Tracking Control Algorithms for Pneumatic Cylinder Actuators[J]. IEEE/ASME TRANSACTIONS ON MECHATRONICS, 2007. 12, NO. 557-561.
[8] De Mengyuan，Tao Guoliang，Zhu Xiaocong, Adaptive Robust Motion Trajectory Tracking Control of Pneumatic Cylinders[J]，Journal of Central South University，20(12),pp3445-3460,2013/1.
[9] De Mengyuan，Tao Guoliang ，Zhu Xiaocong, Integrated Direct/Indirect Adaptive Robust Motion Trajectory Tracking Control of Pneumatic Cylinders[J], International journal of control,86,pp1620-1633,2013/5.
[10] Taehwan Kim and Taeseon Yoon, "Artificial Neural Network Hybrid Algorithm Combimed with Decision Tree and Table," International Journal of Machine Learning and Computing vol.5, no. 6, pp. 471-475, 2015.
[11] Toshio Ito, Supervised Learning Methods of Bilinear Neural Network Systems Using Discrete Data, International Journal of Machine Learning and Computing vol. 6, no. 5, pp. 235-240, 2016.
[12] Petra Vidnerová and Roman Neruda, Kernel Function Tuning for Single-Layer Neural Networks, International Journal of Machine Learning and Computing vol. 8, no. 4, pp. 354-360, 2018.
[13] S. C. Gi, K. L. Han and H. C. Gi, A Study on Tracking Position Control of Pneumatic Actuators using Neural Network[C], Proceeding of 24th Annual Conference on IEEE Industrial Electronics Society, pp. 1749-1753.
[14] Behrad Dehghan, Sasan Taghizadeh, Brian Surgenor, Mohammed Abu-Mallouh, A Novel Adaptive Neural Network Compensator as Applied to Position Control of a Pneumatic System[J],Intelligent Control and Automation, pp388-395, 2011/2.
[15] G. Kothapalli and M. Y. Hassan,Design of a Neural Network Based Intelligent PI Controller for a Pneumatic System[J], IAENG International Journal of Computer Science, 35(2), 2008, pp 217-225
[16] Carneiro J., Almeida F. Reduced-order thermodynamic models for servo-pneumatic actuator chambers[J]. Proceedings of the Institution of Mechanical Engineers, Part I: Journal of Systems and Control Engineering, 2006, 220(4):301–313.
[17] SMAOUI M, BRUN X, THOMASSET D. A study on tracking position control of an electropneumatic system using backstepping design [J]. Control Engineering Practice, 2006, 14: 923-933.
[18] Fangan Meng, Guoliang Tao,Hao Liu. Research on Pneumatic Position Servo Control Strategy and DSP Controller[C], IEEE International Conference on Advanced Intelligent Mechatronics, 2017:279-284.
[19] F.lewis, Neural network control of robot manipulators and nonlinear system[M],1988/8.
[20] Martin Hagan, Howard B.Demuch, Mark Hudson Beale, Orlando De Jesus. Neural Network Design[M].1995/12
[21] Ngoc Tam Bui and Hiroshi Hasegawa, Training Artificial Neural Network Using Modification of Differential Evolution Algorithm, International Journal of Machine Learning and Computing vol. 5, no. 1, pp. 1-6, 2015.

Acknowledgments
This paper is supported by the Fundamental Research Funds for the Central Universities (Project. No. LY13E05004).

IOP Conf. Series: Materials Science and Engineering **575** (2019) 012017 doi:10.1088/1757-899X/575/1/012017

Manufacturing Process Cost Reduction of Agricultural Machinery Part

Ramidayu Yousuk[1,a] and Thidarat Rangsiearam[2]

[1,2]Department of Industrial Engineering, Kasetsart University, Bangkok, Thailand

aCorresponding author's e-mail address: ramidayu.y@ku.th

Abstract. Company ABC is an agricultural machinery part manufacturer. Nowadays demand of such parts are substantially increasing. Cost reduction without affecting product quality become a proper way to raise organization profit. Prior study of manufacturing process of part A shows that the processing cost is too high as a result from short tool life, non-value-added movement of tools, high loading and unloading time, improper number of operator which effect the labour utilization and cost. This project mainly focuses on reducing processing cost by rectifying the problem mentioned above. The improvement includes experimenting to locate the appropriate cutting factors resulted in longer tool life. The shape of drill is changed in order to lengthen tool life. Loading and unloading time is decreased by using the new design of clamp. The movement of cutting tool in machining centre is redesigned and lastly man-machine chart is employed to enhance the labour utilization. As a result, the objective of this project is reached. The processing cost of Part A is decreased from 21.78 Baht to 19.55 Baht or 2.23 Baht which is 10.24% and cycle time also shortens. The proper number of operator is now decreased from 3 operators to 2 operators.

1. Introduction

Nowadays, automotive and agricultural machinery sector are rapidly growing industry among others. This results in high demand of raw material and machinery parts. Entrepreneurs in these industry section are under pressure to produce trifecta products which are high quality, on time delivery and lastly competitive price. In order to lower the price than that offered by the competitors, the production cost is needed to be substantially decreased without affecting product quality. This is the intuition behind this study. Production cost mainly consists of raw material cost, labour cost and overhead cost. One proper way to reduce the production cost and to create competitive advantage is to improve the manufacturing process using engineering techniques and tools such as waste reduction, labour utilization, material selection, and proper tool setting. Prior study of manufacturing process of part A shows that the processing cost is too high as a result from short tool life, non-value-added movement of tools, high loading and unloading time, improper number of operator which effect the labour utilization and cost. These are the seven wastes of lean manufacturing. This paper proposed the application of industrial engineering techniques such as ECRS (eliminate-combine-rearrange-simplify), man-machine chart, redesign of jig and fixture to decrease those 7 wastes which will improve the productivity of manufacturing process. That results in the production cost reduction of Part A at the case study company.

The 7 wastes of lean manufacturing concept is common issue that manufacturing companies face today. It consists with overproduction, waiting, transporting, inappropriate processing, unnecessary Inventory, excess motion, and defects. Many study has been attempt to reduce those wastes. Some of the recent research has been summarized as follow. In [1], the cost reduction was attempted using operations reversal. The results were promising in some case and highly depend on cost parameters and lead times on operations reversal under the cost measure. [2] presented the design and development of a sequence of activities that emphasizes the application of simulation capabilities as a tool to aid the continuous improvement process. A new framework that allows continuous improvement for the reliability of production process and product throughput was presented in [3]. This new framework allows engineers with less effort to define and measure failures of production processes and it includes techniques like Six Sigma DMAIC, FMEA, TOC, FC, swim-line diagram. [4] attempted to reduce the process variation by standardization. It is a powerful tool in improving the organization. Due to space limitation, the detailed of more related work is impossible to present here. However there are many study in this field such as [5,6]. Even though there are different method and concept used in [1-6], the results have led to the process cost reduction. This paper will also focus on reducing the process cost but with different techniques.

ECRS is one of the motion study technique used to improve production lines. It focuses on eliminate unnecessary work, combine operations, rearrange sequence of operations and simplify the necessary operations. In this study, the concept of combine and rearrange will be used in section 4 and 5.

The remaining parts of this paper are organized as follows. Section 2 presents the past research and related work in this field. Section 3 explains the problem description, the manufacturing process and cost analysis of Part A. The cost reduction techniques used in this paper is presented and explains in section 4. Comparative numerical results are also exhibited in this section. Finally, the conclusions are drawn and presented in Section 5 along with future research directions.

2. Problem description

Company ABC produces various kind of agricultural machinery parts. This study mainly focuses only product A which has the total cost of 119.99 baht/piece. The cost structure of product A is shown in Table 1.

Table 1. Cost Structure of Product A.

Cost	Baht	Percentage
Material Cost	77.00	64.17
Process cost	21.78	18.15
Reject cost	2.96	2.47
Administration cost	10.17	8.48
Investment cost	8.07	6.73
Total	119.99	100.00

As shown in Table 1, material cost is the major cost of product A, however, this cost come from the supplier. Therefore, only the process cost is the subject of interest. The objective of company ABC is to reduce the process cost by at least 7% via the improvement of the product A manufacturing process. The process of making product A is divided into 4 operations. As shown in Figure 1, the process cost of operation 1 and 2 is 80.12% of the total process cost. It is worth mentioned that operation 1 and 2 take place in a machining centre with several work elements. Figure 2 shows the process cost of each element in operation 1 and 2.

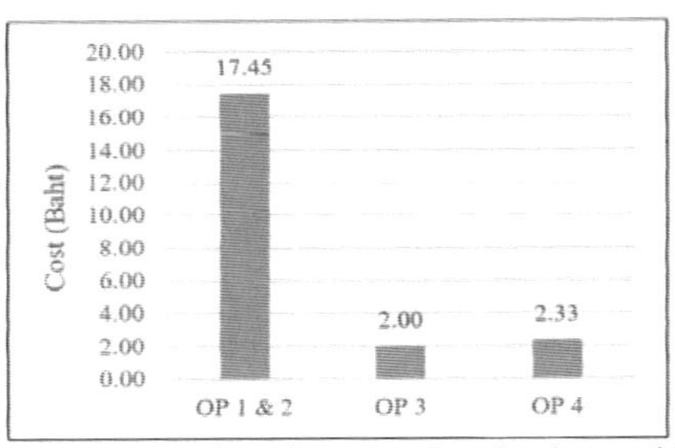

Figure 1. The process cost of each operation in the making of product A.

IOP Conf. Series: Materials Science and Engineering **575** (2019) 012017 doi:10.1088/1757-899X/575/1/012017

Figure 2 shows that rough facing Ø80 mm., finish facing Ø100 mm., and drilling Ø7mm. are the elements with highest process cost. Table 2 presents the process time for each element. Load and unload operation take longer time compared to most work elements. The improvement of process which lead to reduction in process cost is divided into 4 phases,

1. The cutting condition is studied in order to increase tool life or decrease process time.
2. Clamp is designed to improve and reduce the time of the load and unload activity.
3. The transportation waste of tool in Facing Ø80 mm. rough and Facing Ø100 mm. finish are studied and decreased.
4. The waiting waste of operators is studied and decreased.

Table 2. The process time of each element

Operation	Process	Cycle time (sec)	Total time (sec)
Operation 1	Load	29.0	97.5
	Facing Ø80 mm.		
	Rough	17.5	
	Facing Ø100 mm.		
	Finish	17.5	
	Drilling Ø7 mm.	33.5	
Operation 2	Spot facing	6.5	74.5
	Drilling Ø6.8 mm.	7.5	
	Drilling Ø5.0 mm.	5.5	
	Chamfer	8.0	
	Tapping M8x1.25	17.0	
	Tapping M6x1.0	12.0	
	Unload	18.0	
Operation 3	Load	16.0	61.4
	Peeling	32.4	
	Unload	13.0	
Operation 3	Cleaning and		170.1
	Checking	170.1	
	Total		403.5

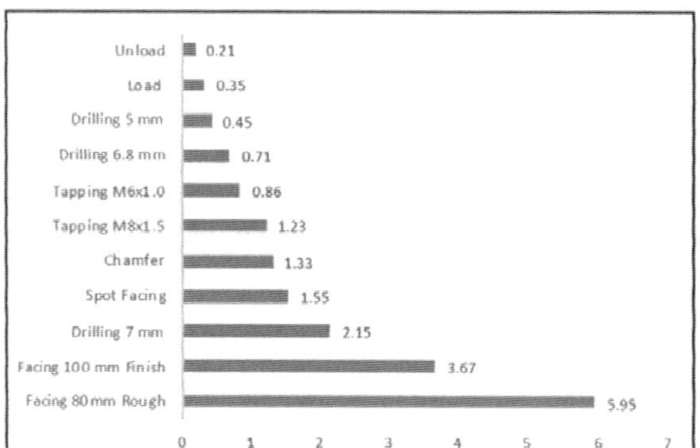

Figure 2. The process cost of each element in operation 1 and 2

3. Improve of cutting condition

Cutting tool cost is one of the major part in process cost [7], in this section, the proper cutting condition is determined in order to extend the tool life which then decrease tool cost. The intuition is that if the company replace the cutting tool too soon it will result in high cutting tool cost. On the contrary, if the cutting tool is not replaced on time, this can cause a process breakdown or defects which can increase labour cost or reject cost. The objective in this section is to lengthen the tool life by determined the appropriate level of cutting factors.

As mention earlier, this study focus only the first three elements with highest process cost which are rough facing Ø80 mm., finish facing Ø100 mm., and drilling Ø7mm. The cutting factors that affect the tool life in these process are cutting speed (m./min.) and feed rate (mm./min.). The experiments are conducted by setting vary the value of cutting speed and feed rate in order to find the optimum cutting condition. The results are summarized in Table 3.

Table 3. The optimum cutting condition

No.	Work Element	Cutting speed (m./min.)	Feed rate (mm./min.)	Tool life (pieces)	
				Before	After
1	Facing Ø80 mm. Rough	120.00	0.20	330	330
2	Facing Ø100 mm. Finish	310.00	0.13	800	720
3	Drilling Ø7 mm.	156.00	1.32	2,000	3,326

As shown in table 3, the tool life is increased for rough facing Ø80 mm., and drilling Ø7mm. but is decreased for finish facing Ø100 mm. However, with the new cutting condition, the process time and the process cost are decreased as shown in Figure 3 and 4.

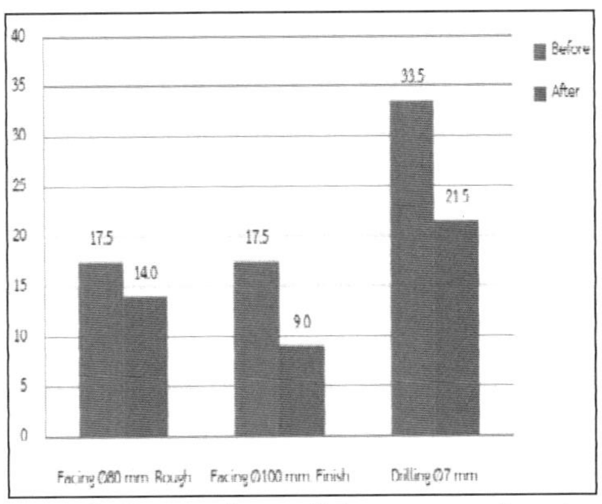

Figure 3. Comparison of the process time (sec.)

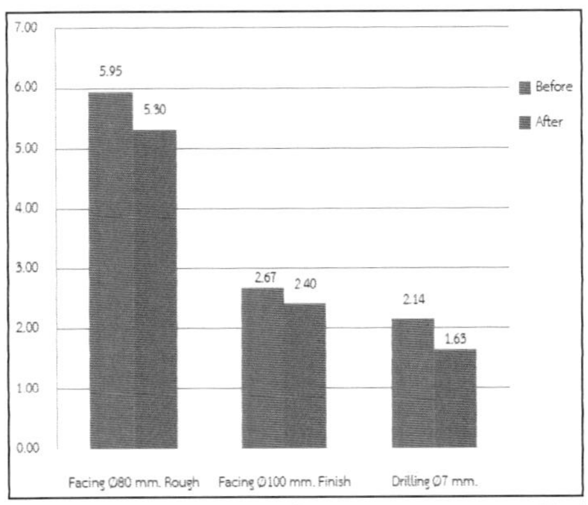

Figure 4. Comparison of the process cost (Baht)

4. Improvement of load and unload activity

Operation 1 and 2 take place in the same machining center. Both operations require clamps to hold the work in a fixed location while the cutting tool operates. Figure 5 demonstrates the location of clamps in operation 1.

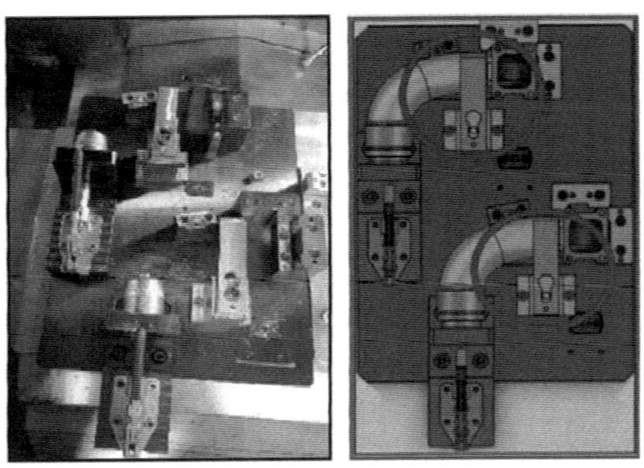

Figure 5. Location of clamps in operation 1

IOP Conf. Series: Materials Science and Engineering **575** (2019) 012017 doi:10.1088/1757-899X/575/1/012017

As shown in Figure 5, two clamps are required in this operation. This results in excessive motion and time in the load and unload activity. The ECRS concept was employed. In the old design, the operator need to operate on the clamps twice each time. The new clamp is designed using SolidWorks in order to improved productivity and reduced time [3]. Figure 6 shows this new clamp and how to use in the machining center. With this new design clamp, the motion has been reduced in half from previous clamp.

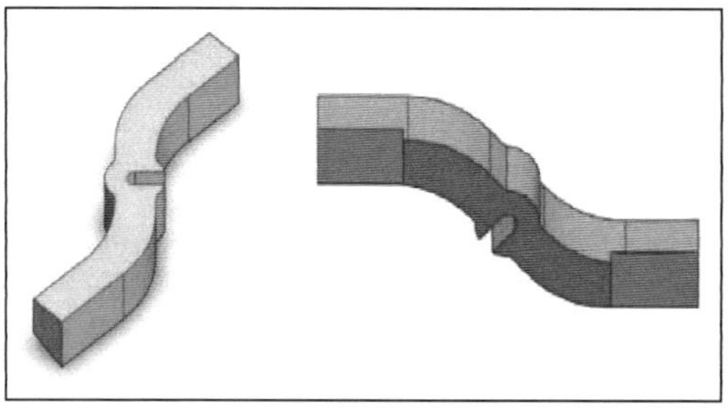

Figure 6. The design of new clamp

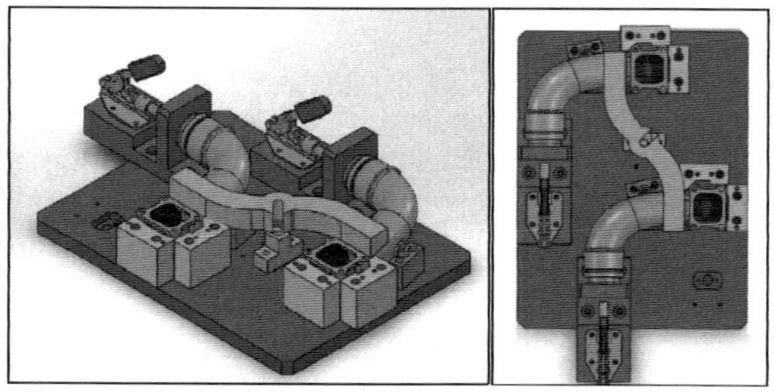

Figure 7. The new clamps in machining center

As shown in Figure 7, by using the new clamp, it combines and simplify load and unload activity which leads to reduction in motion and time. The results of this experiment is are shown in Table 4. It is shown that the time for load and unload activity has been reduced, hence the reduction in process cost. It is worth mentioned that the cost of new clamp is not considered in this study. The material for this new clamp come from the recycle material in the factory.

Table 4. The time and process cost of each element

No.	Work Element	Time (sec.)		Process Cost (Baht)		
		Before	After	Before	After	Diff
1	Load	29.00	22.00	0.35	0.26	0.09
2	Unload	18.00	14.00	0.21	0.17	0.04
	Total	47.00	36.00	0.56	0.43	0.13

5. Improvement of transportation waste

In this section, the cutting tool movements of rough facing Ø80 mm. process and finish facing Ø100 mm. process are analyzed. The result shows many unnecessary moves thus the direction of the cutting tool is redesigned to eliminate these transportation wastes. The concept of rearrange sequence of operations was employed. Figure 8 and 9 demonstrate the cutting tool direction of rough facing Ø80 mm. process before and after the improvement. The improvement result is shown in Table 5. The similar technique is cloned to finish facing Ø100 mm. process. The improvement result is summarized in Figure 10 and Table 6.

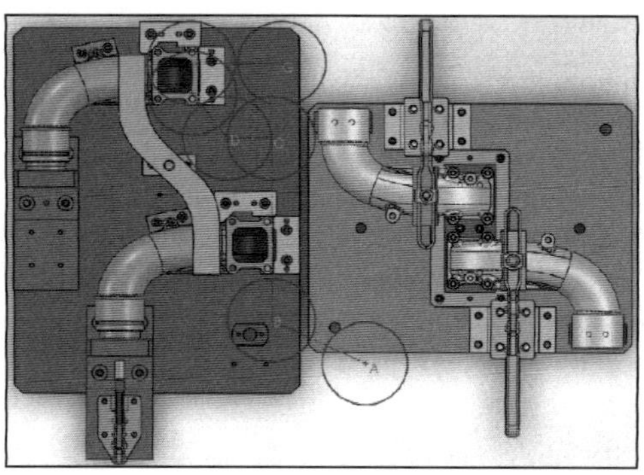

Figure 8. The direction of cutting tool before improvement in rough facing Ø80 mm. process

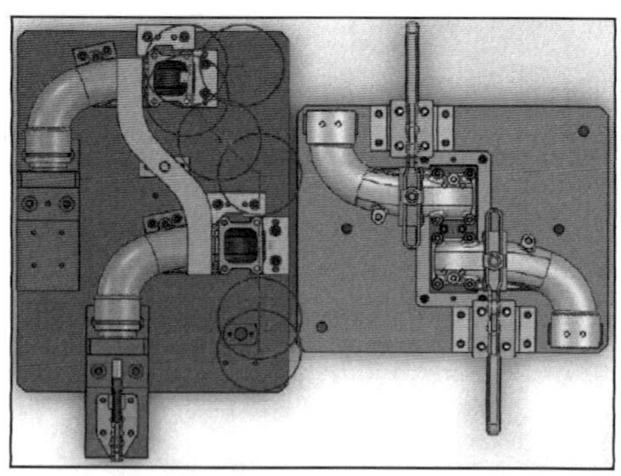

Figure 9. The redesign direction of cutting tool in rough facing Ø80 mm. process

Table 5. Improvement in rough facing Ø80 mm. process

Point	Distance (mm.)		Cycle Time* (Sec.)	
	Before	After	Before	After
A-B	33.00	15.00	0.07	0.03
B-C	34.00	25.50	3.40	3.07
C-D	30.50	15.00	0.07	0.03
D-E	17.75	12.25	1.78	1.48
E-F	14.75	10.00	1.48	1.20
F-G	19.75	9.75	1.98	1.17
Total	149.75	87.50	8.76	6.99

*It is noted that the cycle time depends on distance and table feed.

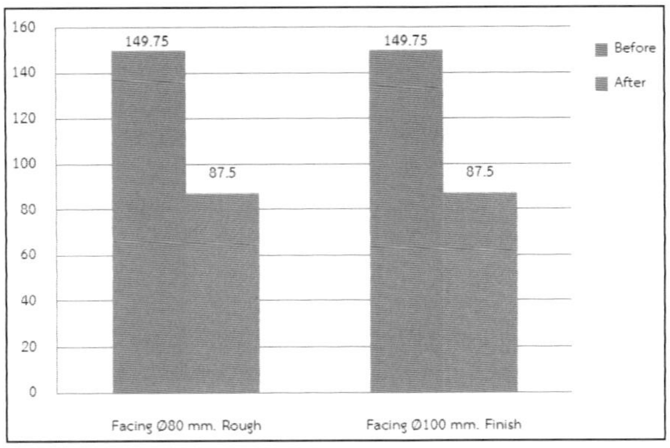

Figure 10. The travel distance reduction of the cutting tool (mm.)

Table 6. The improvement in process cost of each element

No.	Work Element	Time (sec.)		Process Cost (Baht)		
		Before	After	Before	After	Diff
1	Facing Ø80 mm. Rough	8.76	6.99	5.30	5.21	0.09
2	Facing Ø100 mm. Finish	8.76	4.54	2.41	2.18	0.23
					Total	0.32

6. Improvement of waiting waste

The operator in operation 1 and 2 has unnecessary waiting time. Man-machine chart is conducted to confirm this problem. This chart is employed to exemplify the most productive way to use multiple operators, machines or any combination of people and machines [4]. The result is shown in Table 7.

Table 7. Utilization of the operator and machining center in operation 1 and 2

	Operator	Machining center
Idle time (sec.)	80.00	36.00
Working time (sec.)	51.00	95.00
Total time (sec.)	131.00	131.00
Utilization (%)	38.93	72.52

Similar analysis is conduct for operation 3, the result is shown in Table 8.

Table 8. Utilization of the operator and lathe machine in operation 3

	Operator	Lathe machine
Idle time (sec.)	24.40	29.00
Working time (sec.)	37.00	32.40
Total time (sec.)	61.40	61.40
Utilization (%)	60.26	52.77

From the results discussed above, the improvement is proposed. Only one operator is needed to operate both machining center and lathe machine. (One operator for operation 1, 2 and 3) The utilization of the operator and lathe machine after this improvement is demonstrated in Table 9.

Table 9. Utilization of the operator and lathe machine in operation 3

	Operator	Machining center	Lathe machine
Idle time (sec.)	16.40	46.40	92.60
Working time (sec.)	125.00	95.00	48.80
Total time (sec.)	141.40	141.40	141.40
Utilization (%)	88.40	67.19	34.51

The improvement from this improvement is shown in Table 10.

Table 10. Comparison of process time and cost between 1 and 2 operators

No. of operator	Process time of operation 1, 2, and 3 (sec.)	Process Cost (Baht)
2	192.41	17.57
1	141.40	17.22
Difference	51.01	0.35

From the improvement in section 3-6, the process cost reduction in each section is presented in Figure 11.

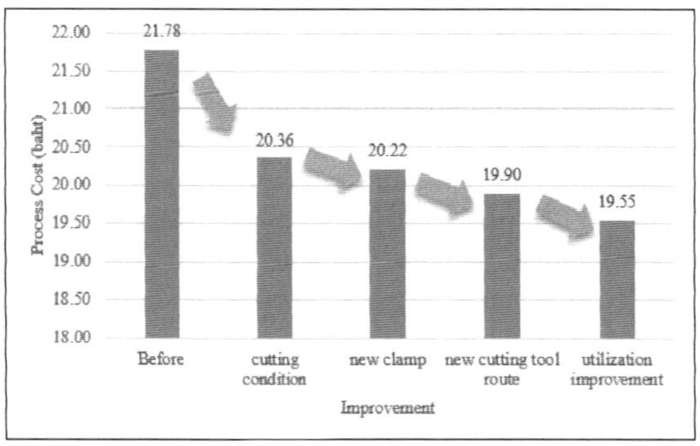

Figure 11. Reduction of process cost

7. Conclusion

This study aims to reduce process cost of agricultural machinery part. Various techniques is employed to improve a manufacturing process such as better cutting tool condition, design of fixture to help eliminate unnecessary motion, the redesign of cutting tool route in machining centre. The result shows the reduction of process cost from 21.78 Baht to 19.55 Baht or 10.24% reduction which is better than the company expectation. The future work should be focused on other work elements in manufacturing process. Some techniques such as simulation, motion economy etc. may be considered.

8. References

[1] Cheung K.L., Song J.S., and Zhang Y., *Eur. J. Oper. Res.* **259**, 1 (2017)
[2] Gomes V.E.O., and Trabasso L. G., *49th CIRP Conference on Manufacturing Systems* (2016)
[3] Jevgeni S., Eduard S., and Roman Z., *25th DAAAM International Symposium on Intelligent Manufacturing and Automation*, (2014)
[4] Mĺkva M., Prajováa V, Yakimovich B., Korshunov A., and Tyurin I., *International Conference on Manufacturing Engineering and Materials* (2016)
[5] Varasquin A., Vieira L. V., and Balbinotti G., *6th International Conference on Applied Human Factors and Ergonomics* (2015)
[6] Holtskog H., *Forty Sixth CIRP Conference on Manufacturing Systems* (2013)
[7] Filho J.C. , Diniz A. E., and Braz J., *Soc. Mech. Sci.* **24**, 1 (2002)

IOP Conf. Series: Materials Science and Engineering **575** (2019) 012018 doi:10.1088/1757-899X/575/1/012018

Research on Antenna Cabin Structure Optimization Based on Response Surface and Multi-island Genetic Algorithm

Sun Weimin[1] **, Cao Weizhong**[1] **, Gu Yeqing**[1]

[1] Nanjing Research Institute of Electronics and Technology,Nanjing 210039, China

Abstract. This paper focuses on the characteristics of high efficiency and high accuracy in the optimization design of the antenna bay structure. The polynomial response surface model is used to replace the computationally intensive target characteristic analysis model in the original genetic algorithm and a multi-island genetic algorithm is combined with the sequential quadratic programming. In this method, the square steel structure of the antenna module is optimized, which not only ensures the search for the global optimal solution, but also improves the calculation efficiency. The optimized results meet the rigid strength requirements of the antenna module under working wind load, and the weight of the antenna module is reduced by 33.5%, which greatly saves costs.

1. Introduction

The antenna module is a carrier and support for antenna arrays, cooling equipment, electronic equipment, etc.; it is an important equipment for ensuring the accuracy of radars and is an important place for on-line repair and maintenance of radar equipment[1]. In order to ensure the accuracy of radar array work, the antenna module needs to meet strict rigidity requirement [2] under the wind load, but at the same time the weight of the module must also meet certain requirements; therefore, the optimization of the structure of the antenna module is very important.

The structural optimization of the antenna bay frame has a large number of design variables. If the calculation of the finite element model is repeated, the cost is too high. At the same time, the multiple design variables cause the complexity and nonlinearity of the optimization process, making it difficult to achieve optimal global optimization results; how to ensure High efficiency and high accuracy in the optimization design are the key to the problem [3].

Therefore, this paper first adopts the method of random super-Latin experiment design to obtain the sensitivity coefficients of each design variable to the weight of the antenna cabin and deformation under wind load; based on this, a polynomial response surface model is established to replace the original calculation model and parallel. Iterative calculations greatly increase the efficiency of optimization. In the optimization problem, a combination of numerical optimization techniques and exploratory optimization techniques is used. The multi-island genetic algorithm is used for global search and the sequential quadratic programming algorithm is used for local search to ensure the accuracy of the calculation.

2. Structural Optimization Design Model

The antenna module acts as a carrier and support for the antenna array and uses a lightweight steel skeleton structure. Viewed from the side, the cross-section is approximately trapezoidal and is divided into 10 layers. The schematic diagram of the antenna tank structure is shown in the Figure 1.

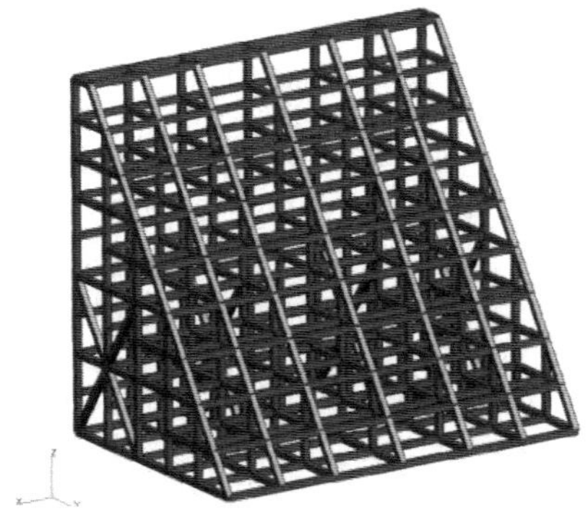

Figure 1. Antenna cabin structure

The skeleton adopts a rectangular steel structure. Three parameters are extracted for each section, which are W, H, and T. Each parameter is shown in Figure 2. According to its hierarchical structure, there are a total of 9 layers and a total of 27 design variables.

Optimize the design model:

$$\text{Min } M(x)$$
$$\text{s.t. } D1(x) \leq Dmax; \ D2(x) \leq Dmax \qquad (1)$$

where M(x) is the weight function of the antenna module, D1(x) is the displacement function under positive blow load, and D2(x) is the displacement function under the side blow load. x=(W1, H1, T1,..., Wi, Hi, Ti,..., W9, H9, T9) denotes the design variable vector, i denotes the ith truss section parameters.

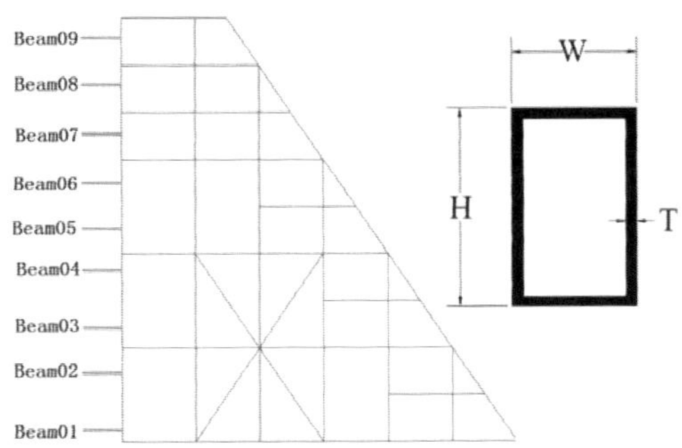

Figure 2. Schematic diagram of design variables

3. Optimization Design Method

Due to too many design variables, experimental design was used to establish a polynomial response surface model. The polynomial response surface model is used instead of the calculation model to overcome the problem of excessive calculation and effective solution to numerical calculation noise. The optimization method uses a multi-island genetic algorithm and a sequential quadratic programming algorithm.

3.1 Test Design Method

The test design is mainly used to determine the design points in the design space that must be tested numerically. In principle, all design variables within the limits of the design variables can be selected; for the sake of efficiency, select a limited number of discrete values (called layers) for each design variable. The test design methods include: central composite test design, full factorial test design, orthogonal test design, uniform test design, and Latin Hypercube test design [4].

The full assay is evaluated at all levels for all combinations of all factors for the experimental design, allowing any number of factors and levels. This method provides a wealth of information for accurately assessing the effects of factors and interactions. There are 27 design variables in the antenna module. Even if only 2 layers are selected for each design variable, the full factorization method needs to calculate 227=1.34×108 times, which is costly to calculate.

Therefore, this paper adopts the random Latin hypercube method, which can greatly reduce the number of sample points. The random Latin hypercube design was the earliest full-space experimental design used in computer simulation experiments. This experimental design is an m × k matrix (m is the number of levels for each variable, k is the number of design variables), each column of data is a random arrangement of m levels, which has an effective space filling capacity; for example, a 2-factor by 9-level study, Figure 3, shows that 81 points are needed for the full factorial, whereas the Latin hypercube design requires only 9 points.

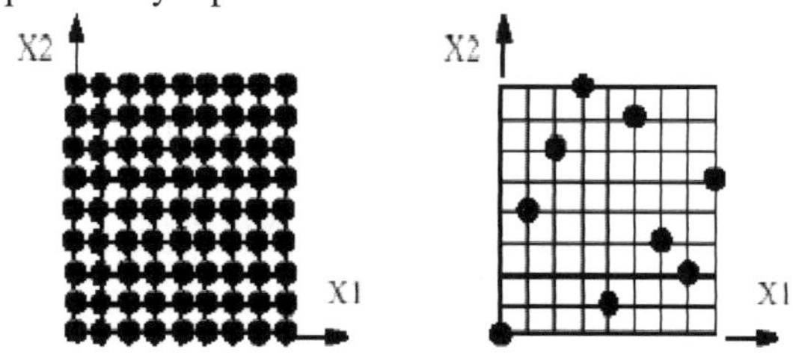

（a）Full analysis due to design（b）Latin super legislation

Figure 3. Full factorial design and latin hypercube design

3.2 Polynomial Response Surface Model Construction

Response surface model is the use of statistics and mathematics knowledge, through a simple expression, the actual analysis of the code for approximation processing to obtain a simpler model, but also conducive to the analysis of the calculation method. The whole process of the response surface method mainly includes: selecting the response surface approximation function model (polynomial model), determining a set of test data points for evaluating the response function, constructing the response approximation function based on the test result, and performing the prediction performance of the response approximation function. Evaluation, final response surface methodology applied to the optimization design [5].

In practical problems, the true response function relationship between input and output cannot be determined in advance, but any function can be approximated by polynomials. Therefore, in common practical problems, polynomial regression can always be used for analysis and calculation regardless of the actual relationship between input and output.

The polynomial response surface model can be written as:

$$y^{(p)} = c_0 + \sum_{1 \le j \le n_v} c_j x_j^{(p)} + \sum_{1 \le j \le k \le n_v} c_{(n_v - 1 + j + k)} x_j^{(p)} x_k^{(p)} \tag{2}$$

Where $y^{(p)}$ is the response of the pth analysis; $x_j^{(p)}$ and $x_k^{(p)}$ are the design values of the pth n_v dimensional design point; $c_0, c_j, c_{(nv-1+j+k)}$ is an unknown underestimated polynomial parameter.

After calculating the underestimated parameters of the response surface equation, it is also necessary to perform statistical tests on the response surface equations to evaluate their approximation to the true response.

The complex correlation coefficient R2 is defined as:

$$R^2 = 1 - \frac{S_{SSE}}{S_{SSY}} = \frac{S_{SSR}}{S_{SSY}} \tag{3}$$

SSSY is the sum of the squared total deviations, SSSR is the regression sum of squares, and SSSE is the sum of the remaining squares. The value of R^2 is between 0 and 1. The larger the value, the more accurate the approximation of the response equation. However, a value of R^2 close to 1 does not necessarily mean a good degree of approximation. This is because the increase in the number of variables in the response equation tends to increase the value of R^2 but does not necessarily increase the prediction accuracy of the equation. Therefore, the modified complex correlation coefficient R^2_{Adj} is defined as:

$$R^2_{Adj} = 1 - \frac{S_{SSE} / (n_s - n_t - 1)}{S_{SSY} / n_s} \tag{4}$$

The approximation degree of the response equation is measured by calculating the value of R^2_{Adj}. If the values of R^2 and R^2_{Adj} are greatly different, then there is likely to be an excess of the variable in the response equation, and if it is approximate, the equation is explained. The more accurate the approximation.

3.3 Optimization Methods

The optimization algorithms in iSIGHT are mainly divided into numerical optimization techniques and exploratory optimization techniques. Numerical optimization techniques usually assume that the parameter space is unimodal, convex, and continuous. Exploratory optimization techniques avoid the search focused on local areas, and these techniques search the global optimal design points across the entire parameter space. The numerical optimization technology can quickly find the local optimum, and the exploratory optimization technology can find the global optimum but the efficiency is low in the later period. Therefore, a search strategy composed of two technologies can take into account both the global optimum and the later efficiency.

Sequential quadratic programming in numerical optimization techniques is used to solve the nonlinear mathematical programming problem with constraints [6]. The main advantage of this algorithm is that it is easy to use with a very robust algorithm. Mathematical convergence and numerical performance are The assumptions of the applicable situation are superior to other mathematical programming algorithms.

Exploratory optimization methods include: simulated annealing algorithm[7] and multi-island genetic algorithm[8].

The simulated annealing algorithm is obtained by simulating the physical process of metal annealing. When the metal is heated to a certain temperature, it melts. At this time, the molecules can move freely. The slow and regular cooling of the metal allows the atoms to be arranged in the most stable and energy-minimized position. Minimizing the energy of the metal, this process minimizes the objective function. The global minimum solution corresponds to the most stable state of matter.

The genetic algorithm mainly uses the law of "survival of the fittest" in the process of biological evolution to imitate the genetic propagation mechanism in the process of biological evolution, and encodes the individual who optimizes the problem space, and then selects, crosses, and mutates the encoded individual population. Iteratively seeks a combination of optimal solutions or better solutions from new populations.

The fitness function is the only criterion for assessing the pros and cons of an individual. The genetic algorithm determines the individual's chance of reproduction based on the degree of fitness. Individuals with a higher fitness value have a greater chance of reproduction than those individuals with a smaller fitness value, resulting in an average new population. The fitness value is higher than the average fitness value of the old group.

The biggest difference between the multi-island genetic method and traditional genetic algorithm is that each population is divided into several subpopulations, also known as islands. The traditional genetic algorithms were performed in the respective sub-populations, and some individuals were selected to “immigrant” to other islands on a periodic basis. Such operations became “immigrants”. There are two parameters that control the immigration process: the immigration interval (the number of progeny to be procreated after each immigration); the immigration rate (percentage of immigration individuals). In this way, populations are separated and more kinds of solutions are generated. The diversity of the solutions improves the ability and convergence of the global search. In order to avoid convergence before the algorithm matures, large groups are used to improve the quality of search, but large groups increase the amount of calculation of individual adaptive evaluation, thereby reducing the convergence rate.

Simulated annealing algorithms and genetic algorithms have many similarities. They all need to generate new design points through mutation from the old design points. The simulated annealing algorithm is simpler than the genetic algorithm because it only checks one design point in the search space each time, and the genetic algorithm checks a set of design points. In addition, the simulated annealing algorithm has fewer parameters than the genetic algorithm. Therefore, the simulated annealing algorithm is very suitable for the discrete design space, while the genetic algorithm can be calculated in parallel.

The optimization problem of the size of the skeleton section of the antenna module should be a multi-modal nonlinear problem. Considering that there are many design variables, the global optimization algorithm is used to explore the multi-island genetic algorithm, and the local search uses the second algorithm in the numerical optimization algorithm. Planning method. This not only guarantees the search for the global optimal solution, but also guarantees the local convergence efficiency. The local optimization adopts the sequential quadratic programming algorithm, and local optimization is performed. At the same time, the finite element model calculation is used to locally update the polynomial response surface model, which ensures the accuracy of the calculation model.

4. Design Optimization

The finite element model was established using Patran, and the model was parameterized by the iSIGHT integrated PCL [9]. Super Latin sampling method was adopted. Polynomial response surface model was established through experimental design. Finally, the optimal algorithm was obtained using a combination of multi-island genetic algorithm and sequential quadratic programming algorithm.

4.1 Finite Element Modeling and Calculation Parameters of Antenna Cabin

The finite element model was established based on the structural dimensions, and the units were in Newton and millimeter units. The finite element model is shown in Figure 1. The antenna module was built using beam elements.

The antenna module is mainly calculated as whether the stiffness meets the accuracy requirement at a wind speed of 25m/s. When the given wind speed is a steady wind speed, the wind pressure calculation formula is (g is the gravity acceleration):

$$q = K_R (K_g K_h)^2 \frac{g}{16} V^2 \quad (5)$$

K_R is the coefficient of wind resistance, which is mainly determined by the shape and direction of the object. Here, when the wind is blowing, it is 1.4. Kg is a gust factor. When the given wind speed is a steady wind speed, the national army standard is 1.42; Kh is a height factor, and it is 1.15. When the 25m/s is a steady wind speed, q=1.429 kN/m 2.

The initial value of the skeleton beam section is taken as W=500mm, H=500mm, and T=10mm.The initial weight M = 481.6 tons.

The design constraints are: positive blow at 25m/s and deformation of the antenna module $\leqslant$ 14.8mm under the side blow load.

4.2 Experimental design

Using the Super-Latin method, 5000 sample points were selected and the numerical range was ±60% of the initial value.

The sensitivity of each input and each other to the deformation of the antenna module under blow load is shown in Figure 4. It can be seen that the thickness of the first layer skeleton steel has the most significant influence on the deformation of the antenna module, reaching 5.33%.

The polynomial response surface equation is constructed based on the calculation results. The approximating degree index of the response surface equation to the true response is: the complex correlation coefficient R and the modified complex correlation coefficient Radj. The response surface equations R^2=0.959 and R^2_{Adj} =0.955 calculated from equations (3) and (4) are all close to 1, indicating that the response surface polynomial fits well with the sample points.

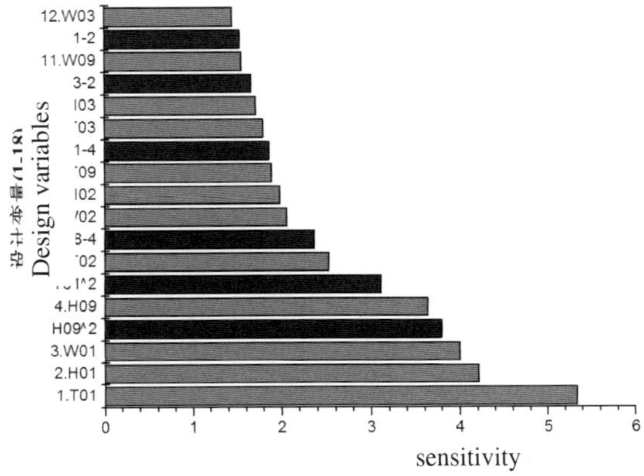

Figure 4. Top 18 sensitivity factors

4.3 Structural Optimization Design Process

Based on the experimental design model, after 43649 iterations, the convergence of the design constraints, optimization goals, and design variables is shown in Figures 5 and 6. It can be seen that in the initial stage of optimization, the global individual population is selected and crossed. Variations and other operations, through iterative search from the new population containing the optimal solution combination, in the latter iteration through the sequential quadratic programming for local search convergence. Since the calculation used is based on the approximate model established by the polynomial response surface and the parallel genetic algorithm is used in the multi-island genetic algorithm, although the number of iterations is more than 40,000, the entire optimization time is only 15 minutes and the calculation efficiency is very high. Figure 5 and Figure 6 show the convergence curves of the deformation of the antenna module under the weight and positive load.

After the final optimized parameters are listed in Table 1, the optimized skeleton weight is 320 tons, and the initial state is 481.6 tons, and the weight loss is 161.6 tons, accounting for 33.5% of the original skeleton weight.

Table 1. List of optimized parameters

Variable Code number	Value /mm	variable Code number	Value /mm	variable Code number	Value /mm
Beam01W	700	Beam04W	207	Beam07W	502
Beam01T	7	Beam04T	7	Beam07T	7
Beam01H	700	Beam04H	290	Beam07H	259
Beam02W	700	Beam05W	207	Beam08W	510
Beam02T	7	Beam05T	7	Beam08T	7
Beam02H	658	Beam05H	235	Beam08H	293
Beam03W	225	Beam06W	361	Beam09W	430
Beam03T	7	Beam06T	7	Beam09T	7
Beam03H	700	Beam06H	260	Beam09H	686

It can be seen from Table 2 that the stiffness of the optimized antenna array has decreased, and the maximum deformation is 14.5 mm, which satisfies the accuracy constraints. After optimization, the maximum bending stress of the antenna frame is -56.2 Mpa, which is based on the stability of the profile pipe. The maximum moment of inertia of the pressure beam I = 8.99 × 10 8 mm 4, area A = 284 mm2, radius of gyration of the profile Mm, L/ρ=11. Conservatively estimate the end support coefficient C = 1, the critical stress curve in reference [10], then the σcr of the rod = 274Mpa, the residual strength coefficient is greater than 2, which meets the stability requirements.

Table 2. Comparison of antenna bay strength before and after optimization

Operating conditions	**Pre-optimization plan**		**Optimized plan**	
	Antenna module deformation/mm	Antenna skeleton strength/Mpa	Antenna module deformation/mm	Antenna skeleton strength/Mpa
Weight	6.63	-36.0	7	-36
25m/s Front	10.6	-47.6	14.2	-56.2
25m/s Side	6.77	-38.3	14.5	-38.3

5. Conclusion

In this paper, a typical antenna module structure is optimized based on response surface and multi-island genetic algorithm. The following conclusions are obtained:

(1) The polynomial response surface model is used instead of the original calculation model. The random super-Latin method is used to carry out the combined sampling of 27 design variables. The exact polynomial response surface model is established through experimental design, and the calculation efficiency is significantly higher.

(2) Taking into account the large number of design variables and the wide design field, in order to avoid focusing on the local area search, an optimization method combining the multi-island genetic algorithm and the sequential quadratic programming algorithm is adopted, thus ensuring the global optimal solution. The search also ensures local convergence efficiency.

Finally, an optimized weight reduction of 33.5% was achieved for the antenna bay frame structure, which will result in significant economic benefits.

References

[1] Shen Wenjun, Shen Qi. Mechanical characteristics and stability analysis of large antenna cabin under wind load [J]. *Mechanical design and research*, 2010, 26 (3): 108-111. (In Chinese)

[2] Zhao Jing. Calculation and test analysis of a skeleton type square cabin [J]. *Electronic machinery Engineering*, 2008, 24 (4): 14-17. (In Chinese)

[3] Cai Xin, Li Hongyue, Wu Yingli, et al. Research progress on optimization design of engineering structure [J]. *Journal of Hehai University*, 2011, 39 (3): 269-276. (In Chinese)

[4] Knill D L, Giunta A A, Baker C A, et al. Response Surface Model Combining Linear and Euler Aerodynamics for Super Sonic Transport Design[J]. *Journal of Aircraft*, 1999, 36(1): 75-86.

[5] Yang Zhouyi, Pong Yongjie, Wang Jian, et. Application of response surface model in multi-objective optimization of boat type [J]. *Journal of Harbin University*, 2011, 32 (4): 407-410. (In Chinese)

[6] Sun Fengcheng, Sun Jian Guo. Engine performance optimization control based on sequence two-time programming algorithm [J]. *Journal of Aeronautical Dynamics*, 2005, 20 (5): 862-867. (In Chinese)

[7] Gu Yuanxian, Xiang Baowei, Zhao Guozhong. Improved simulated annealing algorithm for section optimization design of truss structure [J]. *Journal of Computational Mechanics*, 2006, 23 (5): 546-552. (In Chinese)

[8] Wang Dayang, Zhou Yun. Study on wind vibration control of high-rise structures based on GA optimization theory [J]. *Journal of Civil Engineering*, 2008, 41 (1): 42-47.

[9] Koch P N, Evans J P, Powell D. Interdigitation for Effective Design Space Exploration Using iSIGHT[J]. *Journal of Structural and Multidisciplinary Optimization*, 2002, 23(2): 111-126.

[10] Editorial board of the aircraft Design Manual. Volume Nineth of the aircraft Design Manual [M]. Beijing: Aviation Industry Press, 2001. (In Chinese)

A Low Cost Robotic Medical Simulator for CPR Training

A A Freire Fiallos, F M Rosero Plúa and S N Rea Minango

Departamento de Ciencias de la Energía y Mecánica. Universidad de las Fuerzas Armadas ESPE, Av. General Rumiñahui s/n, Sangolquí, 171-5-231B, Ecuador

snrea@espe.edu.ec

Abstract. Medical simulators have become an important tool for teaching and developing skills in health personnel. The present work describes the design and construction of a simulator for cardiopulmonary resuscitation in adults, of intermediate fidelity, equipped with monitoring software. This device has an adult-size torso, with sensors and actuators controlled and monitored wirelessly through the user's computer. The prototype aims to assist in the training of paramedics and first responders, due to its ability to show real-time data on the execution of the cardiopulmonary resuscitation maneuver. The simulator has been designed in a modular way so that it allows an easy assembly and change of pieces if necessary. The graphical interface gives feedback in real time on the effectiveness of the maneuver. Based on surveys carried out in trained personnel, it was confirmed that the simulator is very useful to carry out practices for which it was designed, as well as to reduce the cost of implementation by being built with materials available in the local market.

1. Introduction

These Cardiopulmonary resuscitation (CPR) is an emergency life-saving technique, used when a person has stopped breathing, or the beating of its heart has stopped. Its purpose is to re-establish regular heart activity, for which chest compressions are administered, enabling air to enter the lungs for oxygenating the blood [1]. CPR maneuver has basic performance aspects, such as proper compression depth and ratio, so that health staff is trained for accurate execution, applying mannequins or simulators.

Simulation is the process of designing a model of a real system and carry out experiences with it, to understand the behavior of the system or evaluate new approaches within the functioning limits, imposed by a certain criterion [2]. For medical simulation, devices and systems have been created and are classified into three levels according to their electronic complexity: a) high fidelity focused on instruction of complex skills; b) medium fidelity, employed to acquire repetitive skills; and, c) low fidelity, for simple skills, designed for heavy use [3].

The use of robotic medical simulators allows the continuous practice of various maneuvers required to save people's lives, where feedback is essential to correct errors and improve technique. However, due to its high cost, its use has been limited in countries like Ecuador, which is why its development is necessary with available, low-cost elements.

2. Materials and Methods

According to [3], there are certain factors that must be considered when designing a CPR simulator to guarantee a CPR maneuver has quality and effectively helps a person:

- The minimum compression rate is 100 beats per minute, and the upper is 120 [4];
- Chest compressions should be applied with a minimum depth of 5 cm in an adult of medium build, and not exceeding 6 cm;
- Avoid leaning on the chest of the patient between compressions;
- Interruptions between compressions should be minimal because they reduce the effectiveness of the maneuver

2.1. *Mechanical structure*

2.1.1. External structure. The base body of the CPR simulator is a plastic mannequin due to the need for a resistant structure for the maneuver execution (figure 1) and it is sized according to the constitution of an average Ecuadorian. For a better representation of the thoracic area and the correct identification of the compression area, models resembling ribs and sternum were created.

The manikin was modified to enable movement in the joints. The union between the body and the head was adapted to allow hyperextension of the neck. The base skull was 3D printed and covered with a latex mask, like the rest of the body, to resemble the human skin.

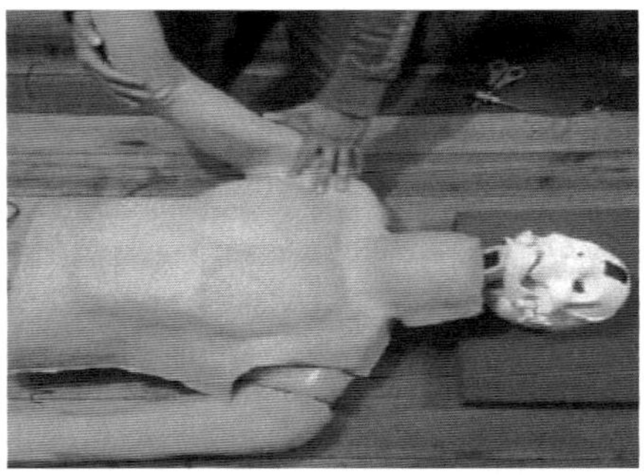

Figure 1. CPR simulator external structure.

2.1.2. CPR structure. The CPR structure that allows performing the compression maneuver (figure 2), is composed by a compression spring, a support table, and the electronic subsystem.

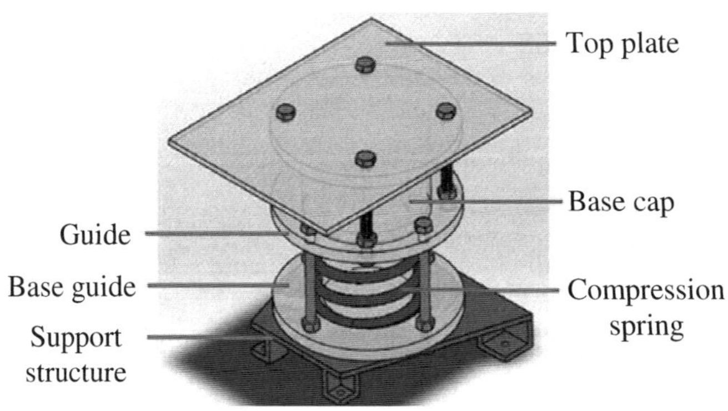

Figure 2. CPR structure.

The compression spring supports the force to move the chest the distance necessary for the maneuver to be correct and was designed to fit in the thoracic cage, using the parameters shown in table 1.

To guarantee a long lifetime and good performance, the spring was designed for a static and cyclic load of 80 kg, although in reality forces during training were rated between 45 and 55 kg.

Table 1. Compression spring parameters.

Parameter	Value
Outer diameter	92 mm
Free length	150 mm
Solid length	70 mm
Wire gauge	8
Material	Stretched solid wire A227
End type	Square frosted

A support table built in ASTM A36 steel give the manikin a flat surface to support the structure and the compressions. The design of this table was validated using CAE software, for an applied force of 80 kg.

For an adequate simulation of the CPR process, it is necessary to emulate the pulmonary behavior, which is why a pulmonary system was created. This system allows 1observing the insufflation of the chest and is composed of a plastic bag of 500 ml (figure 3) placed on a support table and two electrovalves that will control the air passage from a compressor to the lungs or from the airway to them.

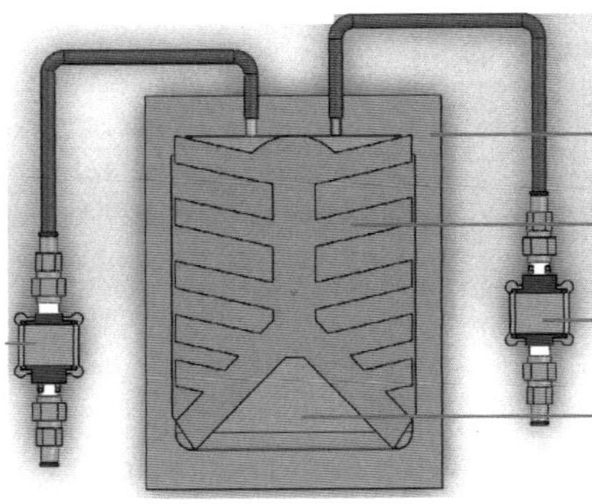

Figure 3. Pulmonary system.

2.2. Electronic system

Two data acquisition boards, a Raspberry Pi 3 model B and an Arduino Nano, were used in the simulator to read both the analogue and digital sensors and are connected to each other through a USB port. Additionally, several sensors and actuators were included in the simulator for specific purposes:

- Proximity sensor - to measure the depth of chest compressions.
- Servomotor – for jaw opening and closing.
- Solenoid valve – to simulate insufflation of lungs.
- Flat vibrating motors – to simulate the carotid, brachial and radial pulse.

Due to a large number of connections generated with the Raspberry and Arduino boards, it was necessary to create another board that was used as a connecting bridge between sensors, elements, and data acquisition cards. The design of the control circuit was based on the need to feed sensors that operate at different voltage values, which are 3 and 5 V. The Raspberry card provides these voltage sources, for this reason, the design connects all the sensor inputs to be used with their respective voltage and to simplify the connections. 12 V were also required to activate the solenoid valves, for this reason, an external source was used. This source also allows the vibrating motors to take the necessary current for their proper operation.

2.3. Software

The software contains one main window, three windows that allow to monitor and control the different processes of the simulator; and two auxiliary windows, which allow observing the results after having performed the CPR maneuver. Its programming was made using Python language.

The main window is used to connect to the control board of the simulator through TCP/IP protocol using the IPV4 address, and to access the windows of the simulator's functionalities.

The window of CPR function (figure 4), allows to monitor and graph the data read by the proximity sensor in real time, to establish the depth of the chest compressions and provide feedback with the results obtained, in the secondary windows.

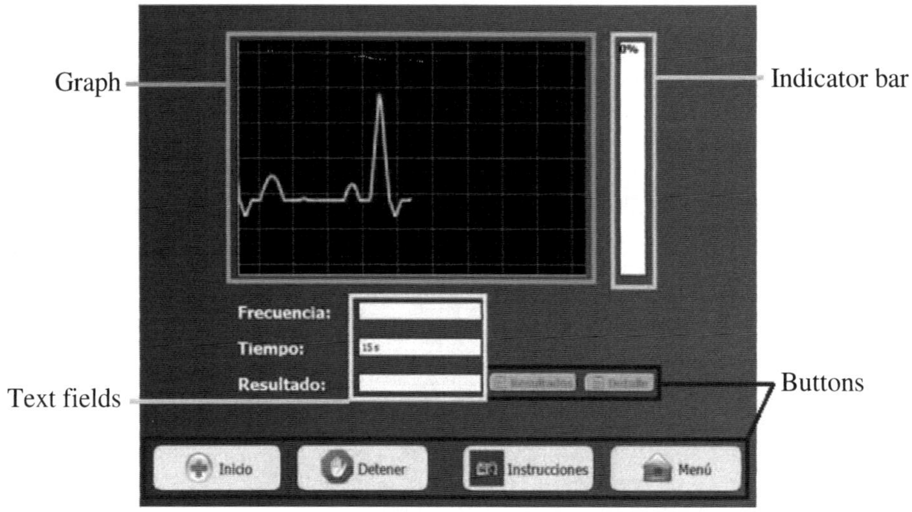

Figure 4. CPR monitoring window.

The software internally has two important modules: one for the CRP training and the other for the different cardiac rhythms simulated. As for the CPR monitoring, the main function of this module is to plot the Sinus rhythm in a time of 1 to 10 seconds, ventricular fibrillation in a time of 10 to 30 seconds and asystole in a time of 30 to 40 seconds. This function detects if there is a compression during ventricular fibrillation and calls the function responsible for taking data from the proximity sensor and plot them in real time. In case of not detecting any compression in the period of ventricular fibrillation, the program will continue until asystole ends.

The second module is capable of representing seven of the most common cardiac rhythms observed in an EKG according to [5], as shown in figure 5. To generate them, an equation composed of six harmonics was used, each harmonic correspond to P, Q, QRS, S, T, and U, based on the work of [6], using the values listed in table 2.

Table 2. Compression spring parameters.

	a	*I*	B	*l*	X
Harmp	0.25	0-100	9.26	0.4167	0.17:0.01:2.17
Harmq	0.25	0-100	12.62	0.4167	0.176:0.01:2176
Harmqrs	1.6	0-100	7.57	0.4167	0.01:0.01:2
Harms	0.25	0-100	12.62	0.4167	-0.08:0.01:1.91
Harmt	0.35	0-100	5.87	0.4167	-0.19:0.01:1.8
Harmu	0.035	0-100	17.5	0.4167	-0.423:0.01:1.567

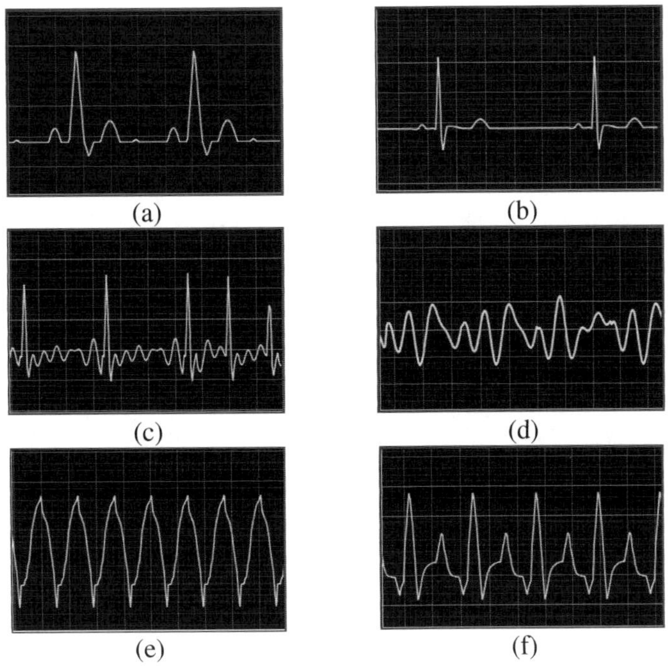

Figure 5. Cardiac rhythms generated. (a) Sinus rhythm; (b) Sinus bradycardia; (c) Atrial fibrillation; (d) Ventricular fibrillation; (e) Ventricular tachycardia; (f) Supraventricular tachycardia.

The amplitude, period and width were modified to generate 7 waves to be able to graph them continuously. This module helps the training of medical personnel in the identification of cardiac arrhythmias.

3. Materials and Methods

For testing the prototype, "RCP" tab was activated and an EKG showing sinus rhythm appeared. When fibrillation starts, chest compressions were applied. After starting the chest compressions, their frequency and intensity are shown in the graphical interface. The graph shows peaks corresponding to each compression made (Figure 6).

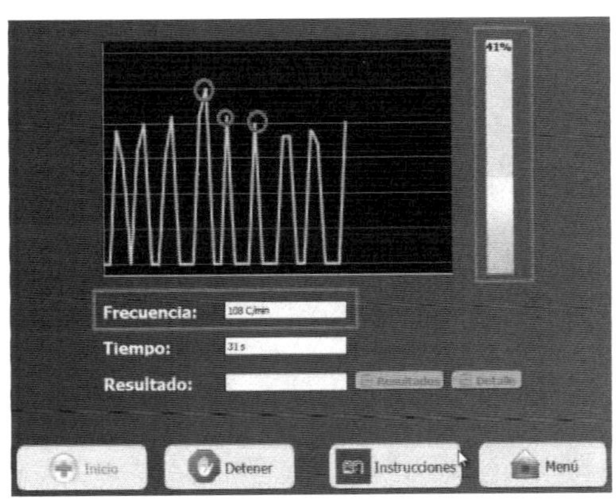

Figure 6. CPR monitoring window showing chest compressions.

A general summary of the practice is shown at the end of the training (see figure 7). When pressing the "Detail" button a list of the executed compressions appeared, classified by color as shown in figure 8.

Figure 7. CPR's results window.

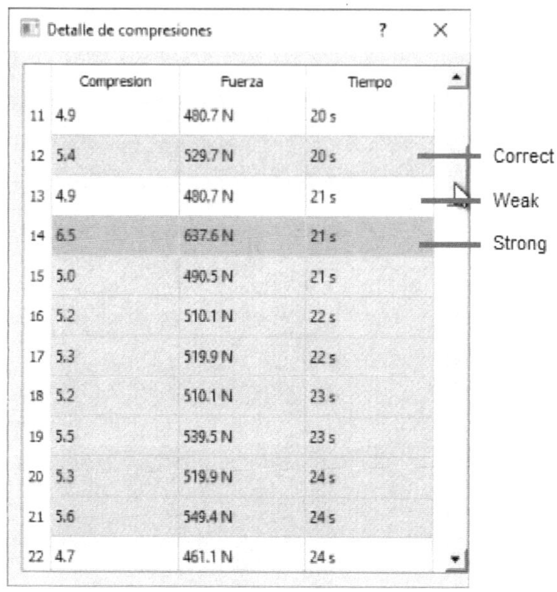

	Compresion	Fuerza	Tiempo
11	4.9	480.7 N	20 s
12	5.4	529.7 N	20 s
13	4.9	480.7 N	21 s
14	6.5	637.6 N	21 s
15	5.0	490.5 N	21 s
16	5.2	510.1 N	22 s
17	5.3	519.9 N	22 s
18	5.2	510.1 N	23 s
19	5.5	539.5 N	23 s
20	5.3	519.9 N	24 s
21	5.6	549.4 N	24 s
22	4.7	461.1 N	24 s

Figure 8. CPR's detailed results window.

4. Discussion

The details and anatomically elaborated parts give the simulator a more realistic appearance, for example, the head allows its inclination and jaw opening for the correct application of insufflations after each cycle of compressions, and this gives realism to the maneuver.

The practices carried out by the personnel of the Ecuadorian Red Cross (figure 9) show that the CPR structure has adequate resistance that simulates the torso of a person, based on the fact that applying a force of 50 kg in a person of normal build generates a displacement of the torso of 5 cm, parameter recommended by the American Heart Association for quality CPR.

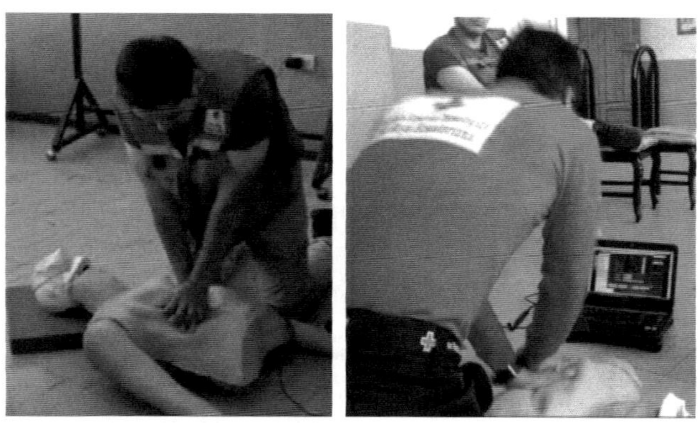

Figure 9. Validation of the prototype.

The user interface facilitates the performance monitoring in the CPR maneuver. The wireless connection between the elements of the simulator and the user interface allows the personnel operating the device to circulate freely in the work area in which the simulator is positioned, in this way the structure of the simulator is not compromised. Additionally, each of the cardiac rhythms loaded in the software presents unique characteristics when graphed.

The results obtained in the tests were positive, but it is important to mention that there are recommendations that can significantly enhance the designed simulator. These recommendations are:

- Expanding the library of cardiac pathologies and adding sounds would be very useful for those who use the simulator;
- CPR interface can be improved, including a scenario where if the maneuver is not being performed properly, the software stops reading data and automatically restarts the execution so that the user detects when his practice is not correct;
- Adding extra functions to offer a more realistic scenario.

Finally, the total value of the project determines that it is possible to produce a low-cost medical simulator, using materials available in Ecuador, with good functionality and anatomical structure. Comparing it with a "MegaCode Kelly Basic" simulator that has similar characteristics, the low-cost prototype represents a 71% savings, proving the viability of the project.

5. Conclusions

The research carried out on the anatomy and dimensions of the different parts of the body such as skull, torso and respiratory system streamlined the design of the structure of the simulator, the selection of sensors, electronic and mechanical devices necessary for the project. It also facilitated the CAD design of the structure to properly position each of the elements used, without compromising the performance of the physical structure of the simulator.

The control and monitoring software developed for the simulator validates the design of the mechanical components that intervene in the CPR maneuver, which together with the sensors generate the necessary information to provide feedback to each of the practices carried out by the students.

The interface of clinical scenarios, being fully configurable, allows students to program different situations that can compromise the health of a patient, thus practicing the actions to be taken when facing those anomalies.

The tests carried out by the staff of the Red Cross, which is trained to perform the CPR maneuver, corroborated the hypothesis because the surveys confirmed the applicability of the simulator implemented in the instruction to personnel of health, as well as its help in the generation of skills.

6. References

[1] AHA 2015 *Aspectos destacados de la actualización de las Guías de la AHA para RCP y ACE*
[2] Galindo J and Visbal L 2007 Simulación, herramienta para la educación médica. *Salud Uninorte* **23** p 79-95
[3] Narváez M 2015 *Ag. Inf. Conacyt* Available: http://www.conacytprensa.mx/index.php/ciencia/salud/2755-centro-de-simulaciones-medicas-implementacion-tecnologica-para-el-aprendizaje-practico.
[4] Mitchell B *Introducción a las arritmias* Available: https://www.msd manuals.com/es-ec/hogar/trastornos-del-corazón-y-los-vasos-sanguíneos/arritmias/introducción-a-las-arritmias
[5] MedLine 2018 *Arritmias* Available: https://medlineplus.gov/spanish/ency/article/001101.htm
[6] Raviprakash K 2016 *ECG Simulation using MatLab* Available: https://www.mathworks.com/matlabcentral/fileexchange/10858-ecg-simulation-using-matlab

Acknowledgments
Support provided by Universidad de las Fuerzas Armadas ESPE was greatly appreciated.

Study on a Real Time Scheduling Optimization Model for Multi-source Heterogeneous Distribution Network

Hongyang Huang[a], Xiaoming Huang, Boliang Lou, Junchao Ma, Haiqing Lu, Wentao Lv

State Grid Zhejiang Electric Power Company Electric Power Reasearch Institude, Hangzhou 310014, Zhejiang Province, China

Corresponding author: 13858091059@139.com

Abstract: Real-time scheduling for distributed power consumption can be heterogeneous distribution network problems, in this paper, an optimization model based on real-time scheduling based on the day before the scheduling model, the power supply capacity of the model based on load forecasting to make full use of distributed power and energy storage equipment, reduces in the peak period of the main power supply rely on, and the imperialist competitive algorithm based on the optimization model was solved. Through the simulation test, the power supply capacity of the real-time scheduling model can effectively utilize the distributed power and energy storage devices to reduce the peak power grid based on the running cost of distribution network is reduced by 2.59%.

1. Introduction

With the development of power generation technologies such as photovoltaic and wind power, the number and capacity of distributed energy connected to the distribution network are rapidly increasing. This makes the network structure of the distribution network complicated to change, and is no longer just a matter of power distribution. Actively dissipate distributed energy with intermittent power supply, and effectively dispatch distributed power and energy storage equipment in the power grid[1]. Therefore, the traditional real-time scheduling model is more and more difficult to deal with the development of multi-source heterogeneous distribution networks and urgently needs to be optimized.

In order to fully optimize and coordinate the dispatching management and efficient use of distributed energy, scheduling for multi-source heterogeneous distribution networks is generally scheduled from both the current and real-time levels[2,3]. At the day-to-day scheduling level, the scheduling model uses load forecasting and information delivery data in the load area as the basis. Global scheduling algorithms are used to derive global scheduling strategies to aggregate, distribute, and control global information and real-time effective information [4]. At the local level, the real-time status of distributed energy in grid-connected networks is mainly adjusted, and the scheduling strategy is adjusted in real time so that the power generation of distributed energy can be optimally absorbed [5,6].

Literature [7] proposed a load forecasting scheme for distribution network power supply capacity. Based on the real-time special detection of distributed energy supply capability, this scheme accurately predicts the growth rate of load, combines the repeated power flow method to construct an expected change curve for the power supply capacity of the distribution network, and optimizes the dispatch for the possible local power supply limitations. The literature [8] takes the real-time electricity price as the scheduling goal, and based on the information exchange of the main network, proposes an optimization algorithm to autonomously schedule the distributed power source and energy storage equipment, and realize the efficient use of electric energy of the distribution network. Literature [9] proposes the scheduling based on multiple autonomous regions and consensus algorithms for the disadvantages of traditional scheduling algorithms such as low efficiency, poor adaptability, and difficulty in improving distributed power utilization efficiency in multi-source heterogeneous distribution networks. The strategy improves the operation speed of the ground algorithm and reduces the amount of system communication data, and it responds well to changes in the structure of the distribution network. In [10], more and more inter-regional power support will occur in the automatic power generation control (AGC) process caused by grid-connected intermittent energy sources, and the problem will be modeled and solved by differential game theory. Make the game reach the non-cooperative feedback Nash equilibrium point. The simulation on the two-region AGC model proves that the proposed coordinated control strategy can be guaranteed to be executed voluntarily and faithfully by each region, because the change of strategy alone will lead to the decline of revenue in each region. In [11], aiming at the randomness and low prediction accuracy of wind power, a coordinated economic dispatching strategy for power systems with large-scale wind power is proposed, and a coordinated economic dispatch model for wind power systems based on chance constrained programming is established.

The above literature has proposed a variety of strategies for global-level scheduling optimization for multi-source heterogeneous distribution networks, but there are few analytical studies for real-time scheduling. Based on load forecasting and power supply capacity analysis of distributed power supply, this paper constructs a multi-source heterogeneous distribution network real-time scheduling optimization model, solves the real-time scheduling optimization model through an improved competition algorithm, and finally optimizes the application of real-time scheduling through simulation experiments. The simulation of the model and the improved competitive algorithm is validated.

2. Real-time Scheduling Optimization Model

2.1 Real-time Schedule Model

Before the date of dispatch is the basis for real-time scheduling, providing basic constraints for real-time scheduling. Multi-source heterogeneous distribution network dispatch system block diagram shown in Figure 1.

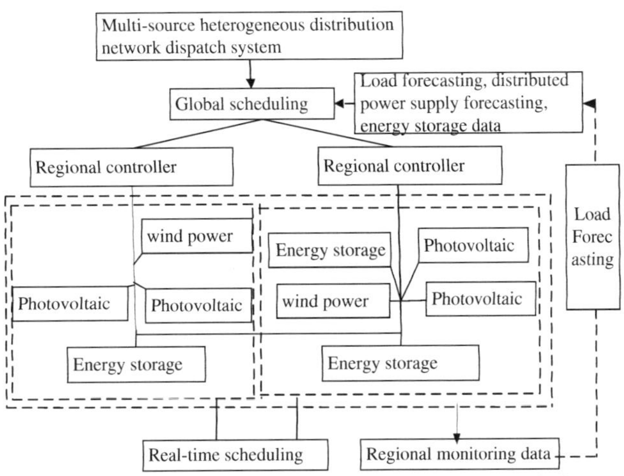

Figure 1. Multi-source heterogeneous distribution network scheduling system structure.

The purpose of the day-to-day scheduling is to construct a scheduling model based on load forecasting, forecasting the intermittent power supply capacity of distributed power supplies, and detection of energy storage data, and to optimize the operation of the distributed power supply to achieve the lowest system operating cost. The ideal power supply quality. The operating costs of the distribution system include the cost of the distributed power supply, the main network purchase cost, the line loss cost, and the energy storage cost [12].

The model of the day-to-day schedule is shown in Equation 1.

$$\min(P_{cost}) = \sum_{t=1}^{T}\left[\sum_{i=1}^{n} c_i(t)P_{1out}^{i}(t)\Delta t + \sum_{j=1}^{m} c_j(t)P_{dg}^{j}(t)\Delta t\right] + \sum_{t-1}^{T}\left[\sum_{k=1}^{H} \Delta C_{ess}(t)P_{ess}^{k}(t)\Delta t + \sum_{i=1}^{n} c_i(t)\Delta P_{1out}^{i}(t)\right] \quad (1)$$

Among them, ci(t) represents the power purchase cost of I on the t time point, and cj(t) represents the operating cost of the distributed generation unit j at the t point of time, ΔC_{ess} indicating the difference between the charge and discharge costs of the energy storage equipment at the t time point. N is the total number of feeders in the distribution network area, and M is the total number of distributed power sources in the distribution network area. $P_{dg}^{j}(t)$ instantaneous power of the distributed power J at the time point of the T is expressed as the consumption power of the line I on the t point of time, $P_{ess}^{k}(t)$ indicating the charge and discharge power of the energy storage device K at the time point of t. $\Delta P_{1out}^{i}(t)$ represents the line loss value at t time point, and its formula is shown in formula 2.

$$\Delta P_{1out}^{i}(t) = \frac{\delta_p(t)}{\delta_p(t) - 1} P_{1out}^{i}(t) \quad (2)$$

Type 2, $\delta_p(t)$ indicates the comprehensive line loss value of the distribution network on the t time point, and the unit is kW.

From Formula 1, it can be seen that the goal of the current scheduling is to minimize the operating costs of the distribution network. When the running cost of distributed power supply is lower than the cost of main power purchase, it is necessary to make full use of the power provided by distributed power supply. When the operating cost of distributed power supply is higher than the cost of main power purchase, the operation of distributed power supply must be reduced. The energy storage unit in the distribution network can temporarily store electrical energy in the event of a power outage, reducing the cost of electricity purchase during peak periods.

2.2 Real-time scheduling optimization model

Real-time scheduling is the real-time adjustment of the daily scheduling. Its main purposes are: 1. Based on the monitoring of the real-time running status of the distributed power supply, according to the current scheduling strategy, the power supply of the distributed power supply is adjusted; 2. The real-time operation of the energy storage unit is performed. State monitoring, based on the current scheduling strategy, switches the charge and discharge status of the energy storage unit in real time and adjusts the charge and discharge power. The data base for real-time scheduling is the prediction of load changes in the short term and the real-time status of distributed power and energy storage units within the distribution network [13].

$$\min(P_{DG}) = \sum_{t=1}^{T}\left[\sum_{i=1}^{n} P_{DG}^{i}(t)c_i(t) + \sum_{j=1}^{m} P_s^{j}(t)c_s(t) + \sum_{k=1}^{o} P_w^{o}(t)c_w(t)\right] \quad (3)$$

Type 3, $P_{DG}^{i}(t)$,he power function of the Chinese controlled distributed power supply, $P_s^{j}(t)$is the power function of the photovoltaic power supply. $P_w^{o}(t)$ is the power function of the wind power supply. I, J and o are the total number of various types of distributed power supply in the distribution network.

(2) the scheduling model of energy storage unit

The expected goal of the scheduling model of the energy storage unit is to minimize the operating cost of the full energy storage unit.

$$\min(P_{ESS}) = \left| \sum_{k=1}^{n} \Delta c_{ess}(t) P_k(t) \Delta T - \sum_{t=1}^{T} \sum_{k=1}^{n} \Delta c_{ess}^{'}(t) P_k^{'}(t) \Delta t \right| \tag{4}$$

$\sum_{k=1}^{n} \Delta c_{ess}(t) P_k(t) \Delta T$ instantaneous operating cost of the energy storage terminal in the 4 Chinese daily scheduling model, $\sum_{t=1}^{T} \sum_{k=1}^{n} \Delta c_{ess}^{'}(t) P_k^{'}(t) \Delta t$ is the instantaneous running cost of the storage terminal after real-time adjustment.

The real-time adjustment of the distributed power supply and energy storage unit needs to meet some basic preconditions.

First of all, the real-time adjustment of the energy storage unit requires the following conditions.

$$\begin{cases} 0 \le P_{k,discharge}^{'}(t) \le \max(P_{discharge}) \\ 0 \le P_{charge}^{'} \le P_{charge} \\ \min(P_k(t)) \le P_k^{'}(t) \le \max(P_k(t)) \end{cases} \tag{5}$$

Second, the real-time adjustment of distributed power supply needs to meet the following conditions.

$$\begin{cases} \min(P_{DG}) \le P_{DG,i}^{'}(t) \le \max(P_{DG}) \\ \min(P_s) \le P_{s,i}^{'}(t) \le \max(P_s) \quad P_{1out}^{i}(t) + P_{DG}^{'}(t) + P_s^{'}(t) + P_w^{'}(t) + P_{ESS}^{'}(t) = \tilde{P}(t) \\ \min(P_w) \le P_{w,i}^{'}(t) \le \max(P_w) \end{cases} \tag{6}$$

The instantaneous total load and instantaneous power supply are equal in the multi-source heterogeneous distribution network, and the expression is as follows.

$$P_{1out}^{i}(t) + P_{DG}^{'}(t) + P_s^{'}(t) + P_w^{'}(t) + P_{ESS}^{'}(t) = \tilde{P}(t) \tag{7}$$

3. Solution of the Model

3.1 Empire competition algorithm

Imperial Competition algorithm a group-inspired optimization algorithm has good convergence speed and has outstanding advantages for low dimensional target optimization [14]. But the Imperial competition algorithm in the operation of the lack of effective information interaction between the Empire, the Imperial forces did not fully consider in the iterative process, but the weakest colony to the strongest empire, resulting in the algorithm easily fall into the local optimal [15]. In this paper, the evolutionary coefficients are introduced to improve the effectiveness of the algorithm in the low dimensional target function. The specific evolutionary steps are as follows.

First, each colony was mutated in the form of 8.

$$CO = CO_1 + \lambda(CO_2 + CO_3) \tag{8}$$

Among them, the 3 random colonies, λ is the coefficient of variation.

Second, cross-operate each target according to type 9.

$$T_i = \begin{cases} CO & rand < \mu \\ CO_i & rand \ge \mu \end{cases} \tag{9}$$

These are the cross coefficients, Rand is the random factor, and Rand's range of values is [0,1].

Finally, according to the greedy strategy, the colony was updated, that is, when the new colonies were more powerful than the old colonies, the Old colony's position was updated.

3.2 The solution process based on imperial competition algorithm

The process of solving the real-time scheduling model using the imperial competition algorithm is as follows.

(1) Initialization of national, national cost function, Empire, assimilation coefficient, colony coefficient and other related parameters.

(2) Encode the power of distributed power supply. When the total quantity of distributed power supply in distribution network is a, the power supply code of power is as shown in 10.

$$\dot{P}_{DG} = \begin{bmatrix} P^{DG}_{1,1} & P^{DG}_{1,2} & \cdots & P^{DG}_{1,B} \\ P^{DG}_{2,1} & P^{DG}_{2,2} & \cdots & P^{DG}_{2,B} \\ \cdots & \cdots & & \\ P^{DG}_{a,1} & P^{DG}_{a,2} & \cdots & P^{DG}_{a,B} \end{bmatrix} = \left[P^{DG}_{a,B}\right] \tag{10}$$

$P^{DG}_{a,B}$ represents the instantaneous power of a distributed power in the B of distribution area. Similarly, it is possible to deduce the coding of the power supply of the C energy storage units in the distribution area B and the power supply $\dot{P}_{ESS} = [P^{ESS}_{c,B}]_t$, coding of the main network $\dot{P}_{ESS} = [P^{ESS}_{c,B}]_t$. Therefore, the coding of individuals in A countries can be deduced from the above variables $N = [\dot{P}_{DG}, \dot{P}_{ESS}, \dot{P}_L]_A$.

(3) using the competition algorithm to optimize the solution of the model, the solution process is shown in Figure 2.

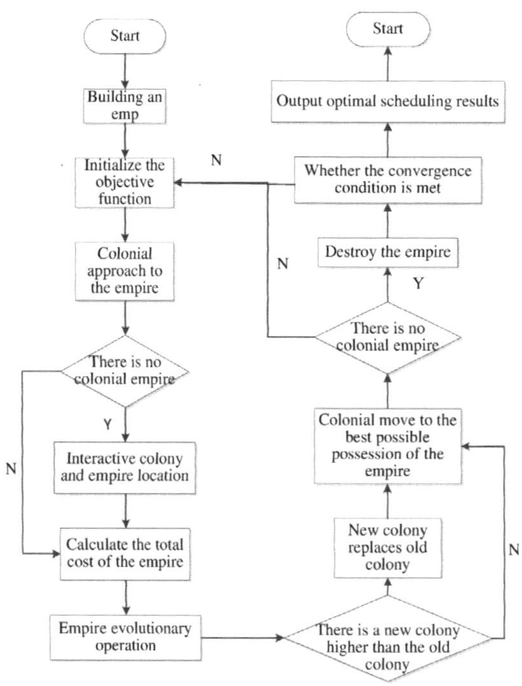

Figure 2. Optimization solution process based on imperial competition algorithm.

4. Simulation Verification

Based on the standard IEEE33 node distribution system, this paper constructs a multi-source heterogeneous distribution system, as shown in Figure 3.

IOP Conf. Series: Materials Science and Engineering **575** (2019) 012020 doi:10.1088/1757-899X/575/1/012020

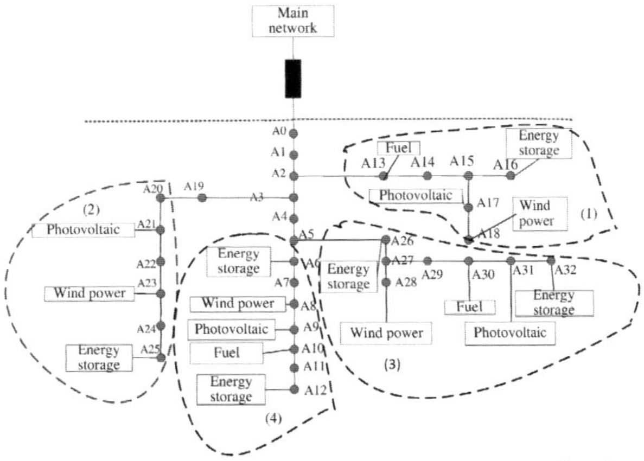

Figure 3. structure diagram of multi-source heterogeneous distribution network

Figure 3 Sets the various types of distributed power and energy storage units to 17 for the larger load nodes, and the configuration of the power and energy storage units is shown in table 1.

Table 1. Distributed power supply parameter configuration table

No.	Node number	Energy type	Installed capacity
1	A6	Energy storage	250KW•h
2	A8	Wind power	500KW
3	A9	Photovoltaic	300KW
4	A10	Fuel oil	300KW
5	A12	Energy storage	250KW•h
6	A13	Fuel oil	300KW
7	A16	Energy storage	250KW•h
8	A17	Photovoltaic	260KW
9	A18	Wind power	500KW
10	A21	Photovoltaic	300KW
11	A23	Wind power	500KW
12	A25	Energy storage	250KW•h
13	A26	Energy storage	250KW•h
14	A28	Wind power	500KW
15	A30	Fuel oil	300KW
16	A31	Photovoltaic	300KW
17	A32	Energy storage	250KW•h

The sales price curve for the main network is shown in Figure 4. It can be seen that during the peak of electricity price, it is necessary to improve the power supply of distributed power supply to reduce the dependence on the main network. During the trough of electricity price, it is necessary to charge the energy storage unit with the main network power supply. Assuming that the fuel generation cost is 0.45 yuan/(kw h), the load of the multi-source heterogeneous distribution network and the power supply of the distributed power supply are forecasted as shown in Figure 5.

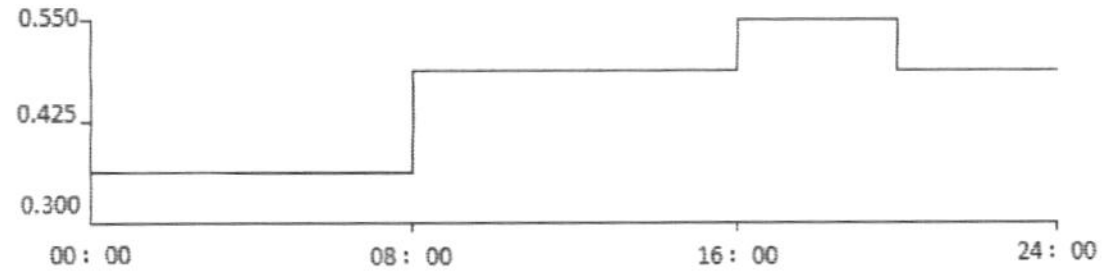

Figure 4. graph of electricity price of main network.

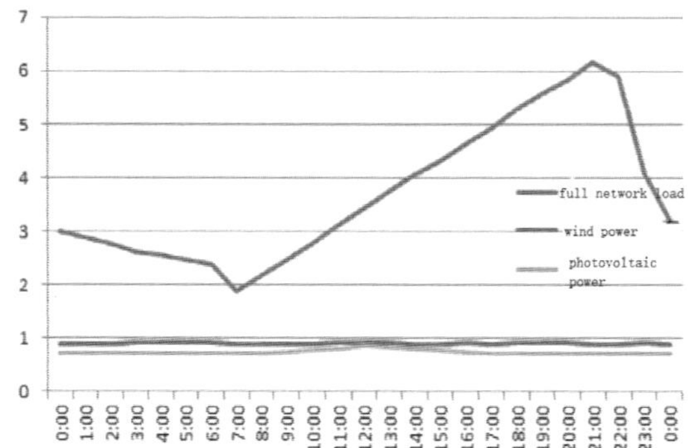

Figure 5. distribution Network Load and distributed power supply power forecast

The Imperial competition algorithm's parameters are shown in table 2.

Table 2. The parameter configuration of the Imperial competition algorithm

reference type	value
maximum number of iterations	200
number of countries	100
number of empire	5
assimilation coefficient	2
offset coefficient	Π/4
colonial coefficient	0.1
coefficient of variation	0.6
cross coefficient	0.9

The simulation test platform is configured for the Intel Xeon e5-2600 v2, 8GBDDR3 RAM and the Windows 10 flagship system, using MATLAB to run the program to solve the scheduling model.

The optimization results of real-time scheduling are shown in Figure 6. According to the current unit operation, combined with the real-time scheduling objective function, the weighted total value of the current system load is 90.646MW, and the system load is reduced by 0.584MW compared with the previous dispatch result of 91.23MW. The real-time scheduling result significantly reduces the peak-to-valley difference of the all-day scheduling, and the system load curve is smoother. In addition, the intermittent distributed energy output is 15.55MW, compared with the 14.95MW of the previous dispatch result, the intermittent distributed energy utilization rate is increased by 3.98%; the gas turbine output is 21.51MW, compared with the 21.34MW of the previous dispatch result, the gas turbine The output increased by 0.8%; the output of the energy storage battery was 2.491 MW, compared with 2.205 MW of the previous dispatch result, the output of the energy storage battery increased by 12.97%; the output of the main network was 53.50 MW, compared with 54.85 MW of the previous dispatch result. Net output decreased by 2.46%. It can be seen from the figure that the utilization rate of distributed energy has been greatly improved during the peak period of electricity price, in which the effective utilization rate of energy storage equipment is maximized, and it can ensure effective charging during the valley period, and participate in system cutting during peak hours. Peak effect; during the valley period, the increase of wind power and photovoltaic output greatly reduced the output of the main network, thus reducing the cost of power purchase. The real-time scheduling cost of the system is 45,876.25 yuan, which is a direct economic benefit of 2.56% compared with the previous scheduled operating cost of 470,83.64 yuan. From the above results, it can be concluded that the system optimization strategy can get the next day system operation strategy. The real-time optimization scheduling can further rationally arrange the distributed energy output according to the current unit operation, which can effectively improve the distributed energy utilization rate and reduce the system operation cost. Improve economic efficiency.

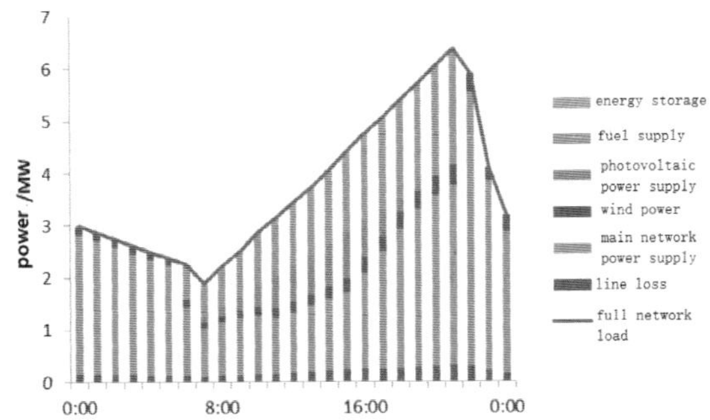

Figure 6. Real-time scheduling optimization results

In order to verify the efficiency of the proposed algorithm in the scheduling optimization problem of distributed energy systems, common ICA, genetic algorithm (GA) and particle swarm optimization (PSO) algorithm are selected as the comparison algorithm. The simulation calculation results are shown in Table 3.

Table 3 Calculative results

algorithm	Number of iterations (real time)	Real-time scheduling of global optimal values/yuans	Real-time scheduling average optimal value/yuan
This article algorithm	35	45876.25	46237.14
ICA	59	47866.63	48161.10
GA	57	46444.96	47000.46
PSO	54	46454.28	46996.00

As shown in Table 3, in the real-time scheduling optimization calculation, the algorithm calculates the optimal value of the system as 45876.25 yuan when the number of iterations is 35. Compared with the other three algorithms, the iteration times of the algorithm are smaller than the common ICA, GA and PSO algorithms, and the optimal values of the algorithm are smaller than the optimal values obtained by the common ICA, GA and PSO algorithms, indicating that the algorithm is distributed energy scheduling. In the application, the solution ability and algorithm adaptability are better, and it can converge quickly in a short time. By improving the optimization method, the algorithm's optimization ability and convergence ability are improved. In addition, compared with the common GA and PSO algorithms, DE-ICA can quickly converge in a short time, so that the optimal value of the algorithm can be obtained. The calculation results fully demonstrate the effectiveness and superiority of the algorithm in the distributed energy system scheduling optimization problem under the active distribution network.

5. Conclusion

Real-time scheduling optimization for multi-source heterogeneous distribution networks The efficiency of distribution network operation can be improved by effective elimination and efficient utilization of distributed energy sources. In this paper, based on the in-depth study of the dispatching model of distribution network, a real-time scheduling optimization model of multi-source heterogeneous distribution network is constructed, and the optimization of the model is solved by the improved Imperial competition algorithm, finally, through the verification of simulation test, the operation cost of distribution network can be reduced by 2.59.

6. References

[1] Guo Tengyun, Liu Yan. Optimization of distribution network reconfiguration with distributed generation based on game method [J/OL]. power system protection and control, 2017,45 (07): 28-34.

[2] Dong Lei, Gong Chengyao, Chen Naishi, Pu Tianjiao. Multi object active distribution network distributed alternating direction method of multipliers optimization scheduling based on [J]. electric power construction, 2017,38 (04): 41-49.

[3] Ding Lei, Liu Junyong, Liu Youbo, Xiang Yue, Su Yun, Zhang Yi. Considering the distribution of PV energy storage capacity of regional distribution grid invested by distributed generation companies [J]. renewable energy, 2017,35 (03): 374-383.

[4] Wang Xiaohui, Chen Naishi, Li Ye, Wang Cunping, Wang Xu, Pu Tianjiao. Trend of linkage of active distribution network optimization scheduling framework of multi-source [J]. power system technology based on 2017,41 (02): 349-354.

[5] Liu Wei, Chen Naishi, Pu Tianjiao, Wang Xiaohui, Dong Lei. The consistency algorithm of active distribution network distributed [J/OL]. scheduling optimization based on Chinese ofthecsee, 2017,37 (06): 1579-1590.

[6] Zhang Hongjun, Wu Yue, Chen Zhuo, Hao Zheng hang. AC / DC distributed power distribution network to accept the real-time simulation of [J/OL]. power system protection and control, 2016,44 (15): 79-85.

[7] Dong Lei, Chen Hui Pu, Chen Naishi, Wang Xiaohui. The great man, control of the active distribution network of multi time scale dynamic [J/OL]. scheduling optimization China ofthecsee prediction model based on 2016,36 (17): 4609-4617.

[8] Zeng Ming, Peng Lilin, Wang Lihua, Li Yuanfei, Cheng Min, sun Chenjun. Active distribution network of distributed energy system with double stage scheduling optimization model [J/OL]. electric power automation equipment, 2016,36 (06): 108-115.

[9] Du Hongwei, Lu Wen, Zhao Junjing, Liu Yalin, Ji Wen Lu city distribution network. The network source load interaction optimization research and application of [J]. scheduling technology of power supply, 2016,33 (01): 45-50.

[10] Jin Yao, Wang Zhengyu, Jiang Chuanwen, Zhao Yan. Changyu, active distribution network and network coordination dispatching [J]. electric power construction, 2016,37 (01): 38-44.

[11] Zhuang Huimin, Xiao Jian. Two stage optimal dispatch model of active distribution network [J]. Journal of Southwest Jiao Tong University, 2015,50 (05): 928-934.

[12] Zhu Jianquan, a Pina, Liu Mingbo. Considering risk and source - Net - real time power load balance scheduling double coordination of [J]. China of the CSEE, 2015,35 (13): 3239-3247.

[13] Wang Wei, Huang Da. Comprehensive optimization of distribution network with adjustable distributed power sources [J]. transactions of Electrotechnical Society, 2015,30 (12): 429-433.

[14] Chen Yunhui, Liu Dong, Ling Wan, Pan Shuchang. Simulation and test platform for coordinated control of active distribution network [J]. automation of electric power systems, 2015,39 (09): 54-60.

[15] Meng Xiaoli, Gao Jun, Gu Wei, Zhu Junpeng, Sheng Wanxing, Wenfei fan. The distribution system with distributed generation before the two stage optimization model [J]. power system technology, 2015,39 (05): 1294-1300.

IOP Conf. Series: Materials Science and Engineering **575** (2019) 012021 doi:10.1088/1757-899X/575/1/012021

Finite Element Model Updating and Verification of Satellite-borne Slot Waveguide Based on Random Vibration Test

Sun Weimin[1] ,Cao Weizhong[1], Tong Mingbo[2]

[1] Nanjing Research Institute of Electronics and Technology,Nanjing 210039, China
[2] Nanjing University of Aeronautics & Astronautics,Nanjing 210016, China

Abstract. For an axial free-floating slot waveguide, finite element model updating and verification based on the vibration data are studied. The slot waveguide and the free-floating side are modeled by shell element and spring element respectively. The two slot waveguide structures are applied to modal and random vibration test getting the modal frequencies, root mean square of acceleration. The finite element are first verified by tested modal data, and then the connecting stiffness will be updated by random vibration test data. These results show that the analytical dynamic responses of the updated finite element are well compared with the test responses, and the maximum error is 3.8%. The spring element could also model the free-floating side well and verify the accuracy for dynamic analysis.

1. Introduction

The slot waveguide is widely used in Satellite-borne SAR antennas[1]. As an important part of SAR, the slot waveguide is complex[2], and its machining accuracy and welding process requirements are strict. The satellite is subjected to complex alternating thermal loads in the orbital operation[3, 4] so that the floating property of slot waveguide must be taken into consideration during installation to avoid thermal stress caused by thermal deformation. At the same time, the antenna bears a large vibration load when the satellite is launching [5], and the antenna deformation caused by the vibration will affect the operation of the satellite to some extent. High rigidity and strength are needed to ensured that the waveguide does not deform when the satellite is working, which contradicts the floating requirement of the waveguide.

Han Jun et al[6] constructed a semi-rigid structural joint unit, which was applied to the model of structural connection parameter identification, and attributed the parameter identification to the nonlinear least squares problem solving. The trust region method was used to solve the problem of too small selection range for initial value of the least squares solution.; an improved Iwan phenomenological model proposed by Wang Dong et al. [7]can better describe the nonlinear characteristics of the joint interface; Ritto et al. [8] applied the Bayesian method to update the probability model with joint stiffness; Lyu et al. [9] identified the bilinear joint stiffness of a semi-rigid joint model from a finite test strain response by using a temperature-based response sensitivity analysis method. Based on the dynamic test model updating method, the joint stiffness of the structure can be effectively identified. Due to the immature of the method to simplify finite element modeling of the floating end of the slot waveguide, the matching between the slot waveguide floating and the stiffness requirement can be determined by the model updating method.

Based on modal parameters, the model updating technique constructs a target residualsis using test data and analysis data to update the model parameters, aiming at obtaining an accurate finite element analysis model[10]. The conventional model updating method focuses on the modification of the structural modes (frequencies and shapes) [11-16], ensuring that the results of structural dynamics of simulation are consistent with the test. On this basis, relevant scholars have investigated on model updating methods for structural response. Canbaloglu et al. [17] used the pseudo-admittance difference method to solve the structural frequency response function, and regarded it as a modified feature quantity to update the structure using model updating method. Wang et al. [18] taken the local connection relationship of structures into consideration and proposed a four-step strategy for updating the connection structure by using the measured frequency response function: localization, characterization, quantization and verification, which can update some simple local-nonlinear structures. Wang et al. [19] used analytical modal decomposition and Hilbert transform to extract transient features in dynamic response data, and update the finite element model with hinged connection structure.

In this paper, two finite element models of a satellite-borne SAR lightweight slot waveguide structure were established under compression and floating case. The simplified analysis model was updated by vibration modes and random vibration test. A finite element modeling method for floating slot waveguide structure was proposed.

2. Satellite-Borne Slot Waveguide Structure

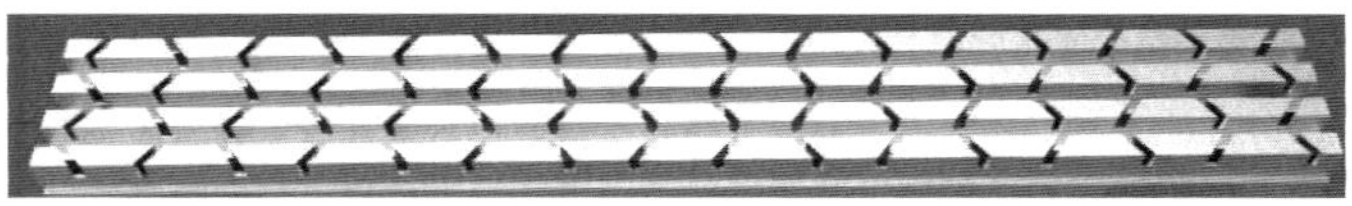

Figure 1. Slot waveguide structure

The slot waveguide is assembled by a plurality of high-precision thin-wall aluminum alloy cavity waveguide components with complex shapes as shown in Figure 1. The thickness of the main body is 1.0mm, and the material is made of aluminum alloy 3003(H12). The material parameters are shown in Table 1. Its geometric size is 296x37x19 mm.

Table 1. Material properties

Type	Elastic Modulus (MPa)	Poisson's ratio	σb (MPa)	σ0.2 (MPa)	σ-1 (MPa)
3003 (H12)	70000	0.33	130	125	55

The slot waveguide is connected to the antenna frame through mounting screws at both ends. In order to reduce the influence of the thermal deformation of the slot waveguide on the flatness of the antenna array in the space environment, the slot waveguide is fixed on one side and free on another side. To achieve zero-friction floating, the floating end screws are installed in a state between compression and relaxation. In order to analyze the dynamic characteristics of this floating state more effectively, the simplified finite element analysis model of the floating slot waveguide is established in two steps and then updated:

(1) The finite element model of the slot waveguide with the floating end compressed was established, and the modal test was performed to verify the accuracy of the corresponding finite element model.

(2) A finite element model of the slot waveguide under floating state was built, and random vibration test of the real structure under floating state was carried out. The model was updated and verified according to this test data.

3. Dynamics Modelingand Testing

3.1 Dynamics model

The slot waveguide finite element model was built by MSC.PATRAN, in which the waveguide cavity take the mid-surface position of each part and is simulated by shell element; the mounting boss and cable mounting flange are simulated by solid elements, the total number of units is 10580, and the finite element model of slot waveguide is shown in Figure 2.

The coordinate system is defined as: x-axis is the slot waveguide length direction (axial direction), y-axis is the slot waveguide width direction, and the z-axis is the slot waveguide vertical direction.

When the floating end is pressed, the movement of the floating end of the slot waveguide in the yoz plane is constrained, so the same constraint boundary condition as the fixed end is adopted.

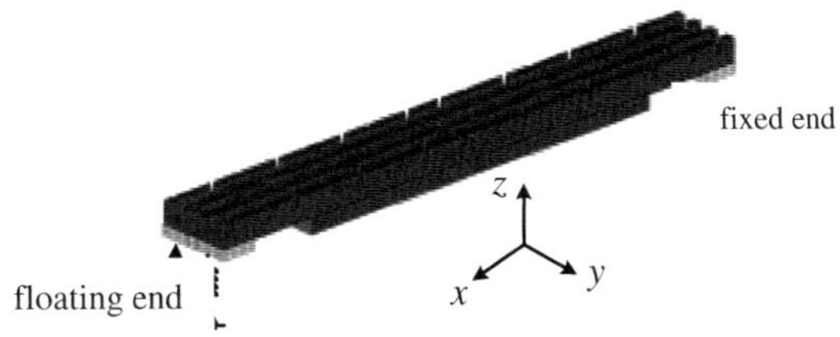

Figure 2. Finite element model of slot waveguide structure

When the floating end is under the free state, the slot waveguide will have a translational motion in the z direction, as shown by the arrow near the floating end in Figure 2. In the finite element model, two spring unit are used at the floating end of the slot waveguide to simulate the connection stiffness of the floating end in y-direction and z-direction. The stiffness in the y direction is infinite, simulating its displacement constraint. The z-direction stiffness is used to simulate the actual floating of the slot waveguide, which needs to be updated according to the vibration test data, and the root of the spring unit is fixed.

4. Finite Element Model Updating Based on Random Vibration

Figure 3 is a schematic diagram showing the process of updating and verification of the finite element model of the slot waveguide based on dynamic characteristics and random vibration. Firstly, the modal analysis of the finite element model under compression is carried out, and the experimental modal frequencies and mode shapes are used to update and verify it. Secondly, on the basis of the verified finite element model, considering the influence of the floating end, a finite element model under floating state is established. The random vibration test data is used to update the spring stiffness at the floating end. Finally, the updated finite element model under floating state is verified by using the random response data.

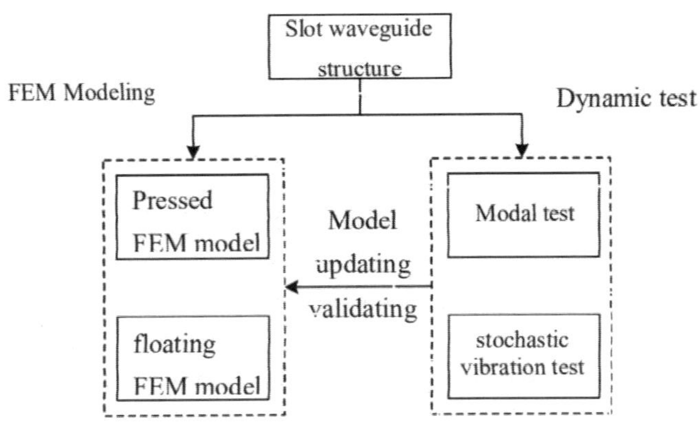

Figure 3. Model updating and validating based on stochastic vibration test

For systems with N degrees of freedom, the equation of motion can be expressed as:

$$\mathbf{M}\&\mathbf{x} + \mathbf{C}\&x + \mathbf{K}\&x = \mathbf{F} \tag{1}$$

M, **C**, and **K** are collectively referred to as structural physical matrices, which are structural mass matrix, damping matrix, and stiffness matrix, respectively. In the modal analysis, the slot waveguide is regarded as an undamped free vibration system, and its characteristic equation is:

$$\left(\mathbf{K}-\omega^2\mathbf{M}\right)\mathbf{x}=\mathbf{0} \tag{2}$$

In order to solve the eigenvalues and eigenvectors, the Block Lanczos method[20] with high precision and calculation speed is adopted. This method with high precision uses recursive algorithm.

Before the random vibration analysis, frequency response analysis of the slot waveguide is firstly carried out to calculate the vibration response of the structure under steady state excitation. The modal method is used for frequency response analysis. Decomposing the dynamic equations using the regular modes of the structure can reduce the scale of the problem and make the solution more efficient. In the frequency response analysis, according to engineering experience, the structural damping coefficient is generally taken as 3%.

When the structure is subjected to a single random excitation, the response power spectral density $\mathbf{S}x(\omega)$ has the following relationship with the excitation power spectral density $\mathbf{S}_F(\omega)$:

$$\mathbf{S}_x(\omega)=\mathbf{H}^*(\omega)\mathbf{S}_F(\omega)\mathbf{H}^T(\omega) \tag{3}$$

$\mathbf{H}(\omega)$ is the frequency response function of the response relative to the excitation and can be obtained from the previous frequency response analysis. $\mathbf{H}^*(\omega)$ is a conjugate matrix of $\mathbf{H}(\omega)$, and $(\cdot)^T$ represents the transpose of the matrix. On this basis, the finite element model of the slot waveguide is updated.

It is known that there are a total of *m* parameters to be updated, and the parameters to be updated change the structural response results by affecting the structural physical matrix. The parameter to be updated can be expressed as:

$$\mathbf{p}=\left[p_1 \quad \cdots \quad p_m\right]^T \tag{4}$$

Then the physical matrix of the structure can be expressed as a function of the variable p to be updated:

$$\begin{aligned}\mathbf{M}&=g_M(\mathbf{p})\\ \mathbf{C}&=g_C(\mathbf{p})\\ \mathbf{K}&=g_K(\mathbf{p})\end{aligned} \tag{5}$$

The corresponding feature quantity can also be expressed as a function of the variable to be updated:

$$\mathbf{f}=F(\mathbf{M},\mathbf{C},\mathbf{K})=F\left(g_M(\mathbf{p}),g_C(\mathbf{p}),g_K(\mathbf{p})\right)=\mathbf{f}(\mathbf{p}) \tag{6}$$

Where f can be any feature quantity, such as modal frequency, mode shape or random response. The finite element model updating can be transformed into a parameter optimization problem:

$$\begin{cases}Min\left\|\mathbf{W}_f J(\mathbf{p})\right\|_2=\left\|\mathbf{W}_f\left(\mathbf{f}_e-\mathbf{f}_a(p)\right)\right\|_2\\ s.t.\ V_{LB}\le p\le V_{UB}\end{cases} \tag{7}$$

Where fe and fa represent the experimental value and the analytical value of the structural feature quantity, respectively, J(p) is called residual term, VLB and VUB represent the upper and lower limits of the change of the structure to be updated respectively , and Wf is the weight matrix among the various feature quantities of the structure.

In general, the feature quantity function f(p) is a nonlinear function of the parameter to be updated. To linearize the nonlinear optimization problem, at the initial updating point, f(p) is expanded to the first-order Taylor expression of the parameter to be updated:

$$f(\mathbf{p})=f(\mathbf{p})+\mathbf{G}\Delta\mathbf{p} \tag{8}$$

Where **p0** is the initial value of the parameter to be updated, and G is the sensitivity matrix, indicating the sensitivity of the feature quantity to the parameter to be updated, which is expressed as:

$$\mathbf{G} = \frac{\partial \mathbf{f}}{\partial \mathbf{p}} = \begin{bmatrix} \frac{\partial f_1}{\partial p_1} & \cdots & \frac{\partial f_1}{\partial p_m} \\ \vdots & \ddots & \vdots \\ \frac{\partial f_n}{\partial p_1} & \cdots & \frac{\partial f_n}{\partial p_m} \end{bmatrix}_{n\times m} \tag{9}$$

Using the Lagrangian multiplier method, the extreme value problem of the equation is transformed into the following linear problem:

$$\mathbf{W}_f \mathbf{G} \Delta \mathbf{p} = \mathbf{W}_f \left(\mathbf{f}_e - \mathbf{f}_a(\mathbf{p})\right) \tag{10}$$

The above equation is a common model updating problem.

4.1 Test content and method

The slot waveguide test pieces are numbered 03# and 05#. Two sets of test pieces were placed on the test bench at the same time. In order to study the vibration characteristics under the compression state and the floating state, the vibration modal test and the random vibration test were performed on the two sets of test pieces respectively.

(1) Vibration modal test under compression state: Identify the modal parameters of the two sets of slot waveguide specimens, and obtain a accurate dynamic analysis model; use PCB-086B80 micro-hammer to excite the test product; Response is measured by the 352C22 micro-acceleration sensor; the VXI dynamic signal analysis system collects and analyzes the excitation signal and the acceleration response signal to obtain the frequency response function of the test piece; Curve fitting and data synthesis of the frequency response function is performed using the I-deas modal parameter identification software to obtain the modal parameters[17, 18] (natural frequencies and mode shapes) of the test piece. Due to the uncertainty of the floating end of the slot waveguide, the floating end of the slot waveguide is pressed by bolts during modal test, eliminating the uncertainty of the model caused by the influence of the floating end.

(2) Random vibration test in floating state: According to the requirements of random vibration test conditions, random vibration tests of three parts in x, y and z directions were carried out, and the acceleration response of the test piece was measured. The floating ends of the two sets of slot waveguide test pieces are crimped by the studs to ensure that their zero-frictional-force floats. Finally, the random vibration response data is used to update the stiffness of the floating end and verify the response.

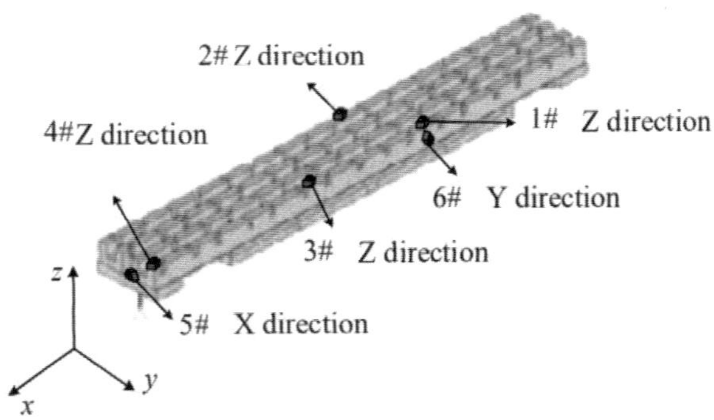

Figure 4. Arrangement of the accelerometers

The slot waveguide was mounted on a test fixture, the test fixture is mounted on a UD-H560B-16 electric vibration table with a 3-point average control. The controller performs random vibration in three directions of x, y, and z according to the test conditions of Table 2.

Acceleration monitoring points are arranged on the slot waveguide. The arrangement are shown in Figure 4. PCB-352C22 miniature acceleration sensor and LDS dynamic signal analyzer are used for acceleration response measurement.

Table 2. Random vibration test condition

Frequency range (Hz)	Acceleration PSD	Totle root mean square of acceleration	Time
20～100	+3dB/oct	14.33g_{rms}	2min
100～600	0.25g^2/Hz		
600～2000	-9dB/oct		

5. Dynamic Model Correction and Verification

5.1 Modal parameter verification

According to the dynamic signal analyzer, the response signal of the acceleration sensor is collected and analyzed, and the frequency response function of the slot waveguide is obtained, as shown in Figure 5.

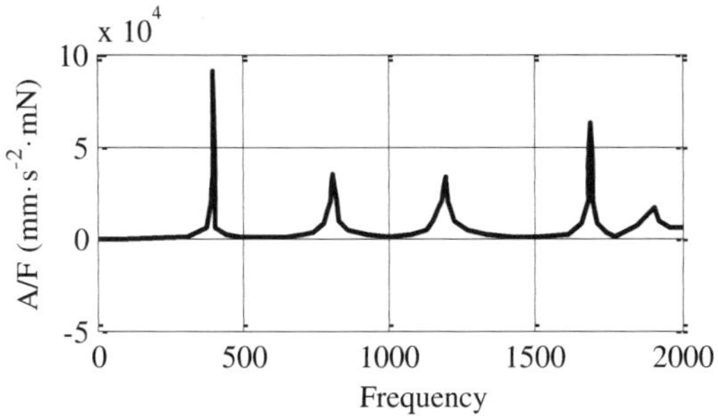

Figure 5. Frequency response function of the slot waveguide

The measured values of the slot waveguide and the calculated values of the finite element model under the compressed state are shown in Table 3. The relative error of the frequency test values between two sets of slot waveguides is small, indicating that the test pieces are consistent.

Table 3. Comparison between tested and numerical modal frequencies

Mode	Test (Hz)		Simulation (Hz)	Error(%)		Shape
	03#	05#		03#	05#	
1st	389.3	394.4	370.9	4.73	5.96	1st Bending
2nd	790.0	813.7	779.98	1.27	4.14	1st Torsion
3rd	1149.0	1197.7	1141.5	0.65	4.69	2nd Bending
4th	1643.6	1689.2	1694.8	3.12	0.33	2nd Bending

It can be seen from the comparison of the above data that the calculation error of the slot waveguide is about 5%, which has a high degree of coincidence. It is indicated that the simplified finite element model of the slot waveguide under compressed state meets the engineering accuracy requirements, which could be employed in the following studies.

5.2 Acceleration response correlation and verification

The acceleration response signals of the stationary segments of each monitoring point are divided into five samples, and the root mean square value of each sample is calculated. The average of the root mean square values of the five samples is shown in Table 4. According to the finite element model under the compressed state, the root mean square value of the acceleration response under the corresponding state is also calculated.

Table 4. Acceleration response of compaction state

No.	Test 1σ(g)	Simulation 1σ(g)	Error(%)
1#	41.8	55.3	32.3
3#	49.3	48.98	0.65
4#	44.5	14.68	67.01
5#	16.6	16.31	1.7
6#	33.6	38.08	13.3

It can be seen from Table 4 that there are three deviations between the calculated values and the measured values of the acceleration monitoring points(1#, 4#, 6#). The 4# monitoring point is close to the floating end of the slot waveguide, and its root mean square value of measured acceleration response in z-direction reaches 44.5g. Since the slot waveguide is designed to ensure frictionless floating, the state at the floating end in z direction is in a non-compressed state. In the finite element model under compressed state, the z-direction freedom of the floating end is constrained. Therefore, the finite element model under compressed state is directly used for calculation, which results in a large deviation between the calculated result and the measured value.

The finite element model under free case is adopted, then the root mean square value of the acceleration response under three different z-direction spring stiffnesses is calculated. The results are shown in Table 5. It can be seen from Table 5 that when the spring stiffness in z direction is 130 N/mm, the deviation between the calculated values and the test results of the acceleration monitoring points is the smallest, and both are less than 4%, which meets the engineering precision requirements. From the above comparison, the spring element could model the free-floating side well and improve the accuracy of dynamic analysis.

Table 5 . Acceleration response of float state

Point	Test 1σ\(g)	UZ=180 N/mm	Error (%)	UZ=140 N/mm	Error(%)	UZ=130 N/mm	Error (%)
1#	41.8	50.12	19.9	44.79	7.15	42.98	2.8
3#	49.3	55.2	11.97	52.35	6.19	51.17	3.8
4#	44.5	36.8	-17.3	43.29	-2.72	45.97	3.3
5#	16.6	/	/	/	/	16.21	2.3
6#	33.6	/	/	/	/	33.49	0.32

6. Conclusion

In this paper, the simplified structural dynamics finite element model of the slot waveguide under compression and floating state is established respectively, and the modal test and random vibration test are carried out on the two sets of test pieces. Based on the vibration modal test, the dynamics model of the waveguide under compressed state is verified. The deviation between the calculated values and the experimental values of the first four-order frequencies is about 5%, which is in good agreement with each other. On this basis, to gain the root mean square value of the acceleration response measured in random vibration, the support stiffness of the slot waveguide under floating state in the z direction is determined. The result shows that the spring element with 130N/mm stiffness in the z direction can simulate the slot waveguide floating effect well. The deviation between the measured value and the calculated value of the acceleration response at each monitoring point is about 3%, which meets the engineering analysis accuracy requirements. This simplified modeling method can be applied to finite element analysis of floating waveguide in the future, a plenty of test costs can be saved.

References

[1] WANG Zhi-gang, YANG Ting-guang and WANG Wei. Structure Design of Microstrip Antenna and Slotted Waveguide Antenna for Satellite-borne SAR[J]. *Electro-Mechanical Engineering*, 2011, 27(3): 40-43. (In Chinese)

[2] WANG Da-zhen, FENG Pei-feng and ZHAO Qin-liang, et al. Study on the Ultraprecision Machining Technology of Aluminum-Matrix Composite Astronautic Parts[J]. *Journal of Astronautics*, 2006, 27(6): 1341-1346. (In Chinese)

[3] MA Hui-tao, LING Jin-dong. Integrated Thermal-structural Analytical Technology for Space Structure with Large-scale and Complicated Construction[J]. *Journal of Astronautics*, 2008, 29(2): 413-419. (In Chinese)

[4] HOU Xin-bin, SHAO Xing-guo, XU Li, et al.Thermal Design and Numerical Results of Chang'e-1 Orbiter [J]. *Spacecraft Engineering*, 2006, 15(4): 21-26. (In Chinese)

[5] SHEN Zhi-chun, LIANG Lu, ZHENG Gang-tie, et al. Vibration Suppression of a Payload Bracket in a Satellite [J]. *Journal of Astronautics*, 2006, 27(3): 503-506. (In Chinese)

[6] HAN Jun, CHEN Huai-hai, NI Hong-wei. Identification of Joint Parameters of Structures Using Measured Frequency Response Function [J]. *Journal of Vibration Engineering*, 2003, 16(3): 355-357. (In Chinese)

[7] WANG Dong, XU Chao, HU Jie, et al. Nonlinear Mechanics Modeling For Joint Interface Of Assembled Structure[J]. *Chinese Journal of Theoretical and Applied Mechanics*, 2018, 50(1): 46-57. (In Chinese)

[8] Ritto T G, Sampaio R, Aguiar R R, Uncertain boundary condition Bayesian identification from experimental data: A case study on a cantilever beam[J]. *Mechanical Systems and Signal Processing*, 2016. 68-69: 176-188.

[9] Lyu M N, Zhu X Q, Yang Q S, Bilinear connection stiffness identification of heritage timber buildings with limited strain measurements [J]. *Engineering.Structures*, 2017. 151:665-681.

[10] Mottershead J E, Link M, Friswell M I, The sensitivity method in finite element model updating: A tutorial[J]. *Mechanical Systems and Signal Processing*, 2011, 25(7): 2275-2296.

[11] CHEN Zhe, HE Huan, CHEN Guo-ping, et al. The research of finite element model updating met hod considering the uncertainty [J]. *Journal of Vibration Engineering*, 2017, 30(06): 921-928. (In Chinese)

[12] Cao Z F, Fei Q G, Jiang D, et al, Substructure-based model updating using residual flexibility mixed-boundary method [J]. *Journal of Mechanical Science and Technology*, 2017, 31(2): 759-769 (In Chinese)

[13] JIANG Dong, FEI Qing-guo and WU Shao-qing. Updating of structural dynamics model with uncertainty based on interval analysis[J]. *Journal of Vibration Engineering*, 2015, 28(03): 352-358. (In Chinese)

[14] WEI Jin-hui, REN Wei-xin. FE model updating based on adaptive response surface method [J]. *Journal of Vibration and Shock*, 2013, 32(08):114-119. (In Chinese)

[15] YU Kai-ping, LIU Rong-he. Model updating of a spacecraft structure based on MRPSO [J]. *Journal of Vibration and Shock*, 2013, 32(17):79-83. (In Chinese)

[16] ZHANG Ming, JIANG Dong, FEI Qing-guo, et al.Finite element model updating using vibration data under base excitation based on sensitivity analysis[J]. *Chinese Journal of Solid Mechanics*, 2011, 32(S1): 280-282. (In Chinese)

[17] Canbaloglu G, Ozguven H N, Model updating of nonlinear structures from measured FRFs[J]. *Mechanical Systems and Signal Processing*, 2016. 80: 282-301.

[18] Wang X, Hill T L, Neild S A, et al., Model updating strategy for structures with localised nonlinearities using frequency response measurements[J]. *Mechanical Systems and Signal Processing*, 2018, 100: 940-961.

[19] Wang Z C, Xin Y and Ren W X, Nonlinear structural joint model updating based on instantaneous characteristics of dynamic responses[J]. *Mechanical Systems and Signal Processing*, 2016, 76–77: 476-496.

[20] Montgomery P L, A Block Lanczos Algorithm for Finding Dependencies over GF(2)[C]. International Conference on Theory and Application of Cryptographic Techniques. Springer-Verlag, 1995:106-120.